THE CITY & GUILDS TEXTBOO[K]

LEVEL 1 DIPLOMA IN
CARPENTRY AND JOINERY

THE CITY & GUILDS TEXTBOOK

LEVEL 1 DIPLOMA IN
CARPENTRY AND JOINERY

COLIN FEARN

MARTIN BURDFIELD

SERIES TECHNICAL EDITOR
MARTIN BURDFIELD

About City & Guilds

City & Guilds is the UK's leading provider of vocational qualifications, offering over 500 awards across a wide range of industries, and progressing from entry level to the highest levels of professional achievement. With over 8500 centres in 100 countries, City & Guilds is recognised by employers worldwide for providing qualifications that offer proof of the skills they need to get the job done.

Equal opportunities

City & Guilds fully supports the principle of equal opportunities and we are committed to satisfying this principle in all our activities and published material. A copy of our equal opportunities policy statement is available on the City & Guilds website.

Copyright

First edition 2013

ISBN 9780851932675

Publisher Fiona McGlade
Development Editor James Hobbs
Production Editor Lauren Heaney

Cover design by Design Deluxe, Bath
Illustrations by Barking Dog Art and Palimpsest Book Production Ltd
Typeset by Palimpsest Book Production Ltd, Falkirk, Stirlingshire
Printed in the UK by Cambrian Printers Ltd

British Library Cataloguing in Publication Data

A catalogue record is available from the British Library.

Publications

For information about or to order City & Guilds support materials, contact 0844 534 0000 or centresupport@cityandguilds.com. You can find more information about the materials we have available at www.cityandguilds.com/publications.

Every effort has been made to ensure that the information contained in this publication is true and correct at the time of going to press. However, City & Guilds' products and services are subject to continuous development and improvement and the right is reserved to change products and services from time to time. City & Guilds cannot accept liability for loss or damage arising from the use of information in this publication.

City & Guilds
1 Giltspur Street
London EC1A 9DD

0844 543 0033

www.cityandguilds.com

publishingfeedback@cityandguilds.com

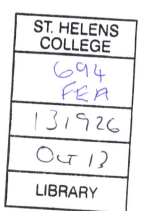

CONTENTS

FOREWORD

Whether in good times or in a difficult job market, I think one of the most important things is for young people to learn a skill. There will always be a demand for talented and skilled individuals who have knowledge and experience. That's why I'm such an avid supporter of vocational training. Vocational courses provide a unique opportunity for young people to learn from people in the industry, who know their trade inside out.

Careers rarely turn out as you plan them. You never know what opportunity is going to come your way. However, my personal experience has shown that if you haven't rigorously learned skills and gained knowledge, you are unlikely to be best placed to capitalise on opportunities that do come your way.

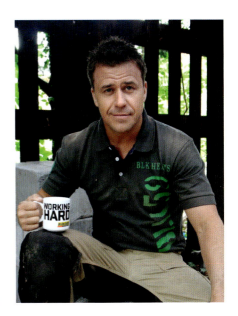

When I left school, I went straight to work in a butcher shop, which was a fantastic experience. It may not be the industry I ended up making my career in, but being in the butchers shop, working my way up to management level and learning from the people around me was something that taught me a lot about business and about the working environment.

Later, once I trained in the construction industry and was embarking on my career as a builder, these commercial principles were vital in my success and helped me to go on to set up my own business. The skills I had learned gave me an advantage and I was therefore able to make the most of my opportunities.

Later still, I could never have imagined that my career would take another turn into television. Of course, I recognise that I have had lucky breaks in my career, but when people say you make your own luck, I think there is definitely more than a grain of truth in that. People often ask me what my most life-changing moment has been, expecting me to say winning the first series of Big Brother. However, I always answer that my most life changing moment was deciding to make the effort to learn the construction skills that I still use every day. That's why I was passionate about helping to set up a construction academy in the North West, helping other people to acquire skills and experience that will stay with them for their whole lives.

After all, an appearance on a reality TV show might have given me a degree of celebrity, but it is the skills that I learned as a builder that have kept me in demand as a presenter of DIY and building shows, and I have always continued to run my construction business. The truth is, you can never predict the way your life will turn out, but if you have learned a skill from experts in the field, you'll always be able to take advantage of the opportunities that come your way.

Craig Phillips
City & Guilds qualified bricklayer, owner of a successful construction business and television presenter of numerous construction and DIY shows

ABOUT THE AUTHORS

COLIN FEARN

CHAPTERS 1–2 AND 4–5

I was born, grew up and continue to live in Cornwall with my wife, three children and Staffordshire bull terrier.

As a qualified carpenter and joiner, I have worked for many years on sites and in several joinery shops.

I won the National Wood Award for joinery work and am also a Fellow of the Institute of Carpenters, holder of the Master Craft certificate and have a BA in Education and Training.

I was until recently a full-time lecturer at Cornwall College, teaching both full-time students and apprentices.

I now work full-time as a writer for construction qualifications, practical assessments, questions and teaching materials for UK and Caribbean qualifications.

In my spare time I enjoy walks, small antiques and 'keeping my hand in' with various building projects.

MARTIN BURDFIELD

SERIES TECHNICAL EDITOR AND CHAPTER 3

I come from a long line of builders and strongly believe that you will find a career in the construction industry a very rewarding one. Be proud of the work you produce, it will be there for others to admire for many years.

As an apprentice I enjoyed acquiring new knowledge and learning new skills. I achieved the C&G Silver Medal for the highest marks in the Advanced Craft Certificate and won the UK's first Gold Medal in Joinery at the World Skills Competition. My career took me on from foreman, to estimator and then works manager with a number of large joinery companies, where I had the privilege of working on some prestigious projects.

Concurrent with this I began working in education. I have now worked in further education for over 35 years enjoying watching learners' skills improve during their training. For 10 years I ran the Skillbuild Joinery competitions and was the UK Training Manager and Chief Expert Elect at the World Skills Competition, training the UK's second Gold Medallist in Joinery.

Working with City & Guilds in various roles over the past 25 years has been very rewarding.

I believe that if you work and study hard that anything is possible.

HOW TO USE THIS TEXTBOOK

Welcome to your City & Guilds Level 1 Diploma in Carpentry and Joinery textbook. It is designed to guide you through your Level 1 qualification and be a useful reference for you throughout your career. Each chapter covers a unit from the 6706 Level 1 qualification, and covers everything you will need to understand in order to complete your written or online tests and prepare for your practical assessments.

Please note that not all of the chapters will cover the learning outcomes in order. They have been put into a logical sequence as used within the industry and to cover all skills and techniques required.

Throughout this textbook you will see the following features:

Ripping

Running with the grain

Useful words – Words in bold in the text are explained in the margin to help your understanding.

INDUSTRY TIP

If using a power tool for the first time, make sure that you have been trained first, and are supervised as appropriate.

Industry tips – Useful hints and tips related to working in the construction industry.

ACTIVITY

Look closely at a pozi-driv and a Philips screw. Sketch the screw heads. What is the difference between the two?

Activities – These are suggested activities for you to complete.

Step by steps – These steps illustrate techniques and procedures that you will need to learn in order to carry out carpentry and joinery tasks.

STEP 1 Loosen holding bolts.

Looking at 'Our House', think about where a chisel would be most useful in building the house. What sort of chisel would you use and why?

Case Study: Karl

Karl has been asked to make the section of cill below:

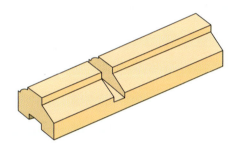

It features a housing joint, a sloping cill, an ovolo moulding and a mortar key. The cill is to be made using power tools using only ready-prepared PAR (planed all round).

- What tools should Karl use to produce the section?
- What additional equipment would be required?

Coping Saw	
	A saw with a thin, narrow blade used for cutting round sharp curves.

At the end of every chapter are some 'Test your knowledge' questions. These questions are designed to test your understanding of what you have learnt in that chapter. This can help with identifying further training or revision needed. You will find the answers at the end of the book.

INTRODUCTION

This book has been written to support students studying carpentry and joinery at Level 1. By studying this book, you should receive a thorough grounding in the skills and knowledge you will need to complete your course and either progress to Level 2, or enter the workforce. You will learn about the wider construction industry and how it works, as well as the skills and techniques you will need in order to work as a carpenter or joiner. You will be able to work safely on site using the correct tools and equipment to produce woodworking joints.

In addition to the features listed on the previous page, which are there to help you retain the information you will need to work with wood, this textbook includes a large trade dictionary. Use this for reference in class and in the workshop. Become familiar with the terms and techniques, and pay attention to the skills you need to master. If you put in the effort, you will be rewarded with a satisfying and successful career in construction.

ACKNOWLEDGEMENTS

I would like to thank my dear wife Helen for her support in writing for this book. My thanks go to the other chaps (Clay, Martin and Mike) for all their help! I dedicate my work to Matt, Tasha and Daisy, and not forgetting Floyd and Mrs Dusty.

Colin Fearn

To my gorgeous wife Clare, without whose constant support, understanding and patience I would not have been able to continue. To Matthew and Eleanor, for not being there on too many occasions, normal service will be resumed. Finally, my parents, to whom I will always be grateful.

Martin Burdfield

City & Guilds would like to sincerely thank the following:

For invaluable carpentry and joinery expertise

Bill Birkin and Steve Redfern.

For their help with photoshoots

Andrew Buckle (photographer), Martin Burdfield, Andrew Kelsey, Robert Fairbairn, Wahidur Rahman and all of the staff at Hackney Community College; Tony Manktelow, Shaun Scofield, Sam Folkes, Charlie Barber and all of the staff at Central Sussex College. Thanks also to Colin Fearn.

For supplying pictures for the book cover

Andrew Buckle.

TRADE DICTIONARY

Industry term	Definition
Abrasive	1 Material used for smoothing wood, includes glass-paper. 2 The wearing nature of a material, for example, rougher surfaces such as chipboard will blunt tools more quickly.
Air seasoning	Natural method of seasoning (drying) timber, where the freshly cut timbers are placed under a cover but air is allowed to pass through the stack. Slowly the timber dries out.
Aperture	An opening, eg in a worktop for a hob to be fitted.
Approved Code of Practice (ACoP)	ACoP gives practical advice for those involved in the construction industry in relation to using machinery safely. The ACoP has a special legal status and employers and employees are expected to work within its guidelines.

Industry term	Definition
Architect	A trained professional who designs structures and represents the client who is building the structure. They are responsible for the production of the working drawings and supervise the construction of buildings and structures.
Architecural technician	A draftsperson who works in an architectural practice. They usually prepare the location drawings for a building.
Architrave	A moulded section that is fixed around a door lining that covers the gap between the lining and the wall.
Asbestos	A naturally occurring mineral that was commonly used for a variety of purposes including: insulation, fire protection, roofing and guttering. It is extremely hazardous and can cause a serious lung disease known as asbestosis.
Auger bit	Rotating cutting tool used in cordless drills and hand braces to bore holes, such as for fitting cylinder and mortice locks.
Backed saw	A saw with either a strip of steel or brass along the top edge that keeps the blade taut. Backed saws are used for the accurate cutting of joints and can be used to cut with and across the grain. *Regional variation: backsaw*
Bass	A traditional tool bag that opens out flat to reveal all of a carpenter's tools.
Belt sander	A portable sanding tool with a power-driven abrasive-coated continuous belt.

Industry term	Definition
Bevel edged chisel 	A chisel with the edges bevelled. Used for light work eg when chopping out the sockets when dovetailing.
Bill of quantities 	Produced by the quantity surveyor and describes everything that is required for the job based on the drawings, specification and schedules. It is sent out to contractors and ensures that all the contractors are pricing for the job using the same information.
Bind	When a saw blade jams because of being stuck in timber. This happens when the timber is damp, not supported properly or twisted.
Block plane 	A small plane that can easily be used with one hand. It has a much lower frog angle improving the finish of the cut.
Blue stain 	A blue fungal discolouration in the sap wood, which does not reduce its strength.
Bond/Bonding 	The arrangement or pattern of laying bricks and blocks to spread the load through the wall, also for strength and appearance.
Boule	A log that has been sawn through-and-through and restacked into the original shape of the log.

Industry term	Definition
Boxed heart	Boxing (a term used in timber conversion) the heart refers to eliminating the heartwood from the boards that would otherwise produce shakes or may even be rotten. This can be achieved by either tangential or radial cutting.
Bridle joint	Similar to a mortice and tenon, in that a tenon is cut on the end of one member and a mortice is cut into the other to accept it. The distinguishing feature is that the tenon and the mortice are cut to the full width of the tenon member.
British Standards Institute (BSI)	The authority that develops and publishes standards for the UK.
Building Regulations	A series of documents that set out legal requirements for the standard of building work.
Butt joint	Simple joint in which two pieces of wood are placed against each other but held with nails, screws, dowels, glue or other fasteners.
Case hardening	A defect caused by timber being kiln dried too rapidly, leaving the outside dry but the centre still wet.
Cavity fixings	Many different types available, used to enable secure fixings when a cavity is present such as stud partition walling.
Cavity wall	Walls built in two separate skins/leaves (usually of different materials) with a void held together by wall ties.

Industry term	Definition
Chop saw	Tool used for making accurate square, angled, and compound cuts (a cut incorporating two angles). Chop saws are on site for a wide variety of tasks including cuts to architraves and skirting, stair spindles, studwork, rafter cuts and any other job where a straight clean cut is required. *Regional variation: mitre saw*
Cill	The horizontal member that bears the upright portion of a frame, ie the bottom part of a door frame. *Regional variation: sill*
Circular saw	A versatile tool used to cut sheet materials.
Combination gauge	Very similar to a mortice gauge but with an additional pin on the opposite side of the stem, allowing it to be used as a marking gauge.
Combination square	A square that measures both 90° and 45° angles.
Concave	Curved inwards.
Concrete	Material made up of cement, sand and stone of varying size and in varying proportions. It is mixed with water.

Industry term	Definition
Conversion	The process of cutting logs into pieces of timber. Four different methods used are boxed heart, tangential cuts, through-and-through and quarter sawn.
Convex	Curved outwards.
Coping saw	A saw with a thin, narrow blade used for cutting round sharp curves.
Countersink	To sink the heads of screws flush with or slightly below the surface of the timber using a countersinking drill bit.
Crosscut saw	A saw used to cut across the board square to the grain, ie cutting across the grain.
Cutting gauge	The same as a marking gauge except that instead of a pin it has a cutting knife and a wedge to hold it in place. The knife severs the fibres of the grain, leaving a clean finish.
Damp proof course (DPC)	A layer or strip of watertight material placed in a joint of a wall to prevent the passage of water. Fixed at a minimum of 150mm above finished ground level.
Damp proof membrane (DPM)	A layer or sheet of watertight material, incorporated into a solid floor to prevent the rise of moisture.

Industry term	Definition
Datum point	A fixed point or height from which to take reference levels. They may be permanent Ordnance bench marks (OBMs) or temporary bench marks (TBMs). The datum point is used to transfer levels across a building site. It represents the finished floor level (FFL) on a dwelling.
Dead load	The self weight of all the materials being used to construct the building.
Defect	Issues with timber, that can either be natural or from seasoning. Shakes and knots are types of defects.
Diamond stone	Used to hone the edges of steel tools, a diamond stone is a plate, sometimes with a plastic or resin base. The plate is coated with diamond grit, an abrasive that will grind metal.
Dovetail joint	Attractive joint with interlocking teeth used for drawers and in fine furniture.
Dovetail saw	Small fine-toothed saw used for cutting dovetails. It is a type of backed saw.
Drilling jig	Drilling jigs are devices which allow repeated processes to be carried out quickly and consistently. An example would be to consistently position door knobs and handles on kitchen cupboards.

Industry term	Definition
Dry rot	Fungal timber decay occurring in poorly ventilated conditions in buildings, resulting in cracking and powdering of the wood.
Dub off	Over-planing the outside edge of timber leaving it thin.
Durability	The resistance of different timber species to fungal and insect attack. Timbers which are resistant are known as durable, timbers that rot easily as known as non-durable.
Edge joint	Edge joints are used to make timber boards wider. They can be 'rubbed' (where glue is applied and the two pieces rubbed together to ensure good coverage of the joint), tongued, biscuited or doweled to increase strength.
Face side and edge	The best two faces of the timber.
Ferrule	A metal band round a wooden handle to prevent splitting.
Finished floor level (FFL)	This is the height of the finished floor level in a property. Represented with a datum point, and horizontal DPC is often installed at this height.
Flat bit	Used for rough boring in wood. They tend to cause splintering when they emerge from the workpiece. They are flat, with a centring point and two cutters.
Floors	The structured layers of a building, eg ground floor, first floor, second floor.

Industry term	Definition
Foundation 	Used to spread the load of a building to the sub-soil.
Friction	1 The rubbing of the saw in the cut. 2 Resistance between the surface of the concrete foundation and the soil around it.
G cramp 	A steel clamp, shaped like the letter C; used to hold, under pressure, two materials placed between the top of the open end of the C and a flattened end of a screw shaft which is threaded through the other end of the C.
Grinding angle 	Angle ground on blade of plane or chisel from which the cutting edge is sharpened.
Grooves	A narrow cut or channel.
Growth ring 	A ring that appears on the inside of a tree, one for every year it has grown. Each ring is split into spring and summer growth.
Gullet 	The gap between the teeth of a saw. These store the waste timber (sawdust) until the teeth are exposed from the timber they are cutting.
Halving joint 	A joint where half of each of the two boards being joined is removed, so that the two boards join together flush with one another.
Hardcore	A mixture of crushed stone and sand laid and compacted to give a good base for the concrete.
Hard point	Where the teeth of a saw, particularly panel saws, are specially hardened during manufacture.

Industry term	Definition
Hardwood 	Timber that has been cut from deciduous trees (broad-leaved trees which lose their leaves in autumn).
Hatch	Marking waste timber with diagonal lines. This prevents mistakes, as only the hatched area is removed.
Haunch 	A shortened stub portion of a haunched tenon.
Heading joint 	A joint between two pieces of timber which are joined in a straight line, end to end.
Health and Safety Executive (HSE)	Government body which provides advice on safety and publishes booklets and information sheets including the ACoP.
High speed steel (HSS)	Material used to make planer knives, cutters, saw blades and router bits. Not as hard or as brittle as carbide.
Honing 	To provide the final, durable polished finish to an edge tool after grinding.
Housing joint 	Joint consisting of a groove usually cut across the grain into which the end of another member is housed or fitted to form the joint, for example stair treads into the string. *Regional variation: trenching joint*

Industry term	Definition
Hypotenuse	The longest side of a right-angled triangle. It is always opposite the right angle.
Imposed load	Additional loads that may be placed on the structure, eg people, furniture, wind and snow.
Improvement notice	Issued by an HSE or local authority inspector to formally notify a company that improvements are needed to the way it is working.
Insulation	Materials used to retain heat and improve the thermal value of the building. Can also be used in managing sound transfer.
Jack plane	A large plane for removing large quantities of material, as in straightening surfaces or reducing the thickness.
Jamb	The upright side member of a door or window frame.
Jigsaw	A versatile power tool used to cut shaped work, ie work that involves curved lines, such as apertures.
Kerf	The width of a cut.

Industry term	Definition
Kiln dried  	Timber dried in controlled conditions in a large oven called a kiln.
Kinetic lifting  	A method of lifting that ensures the risk of injury is reduced.
Laminated  	A plastic sheet glued on top of a base layer like chipboard, as used in kitchen worktops.
Leaves 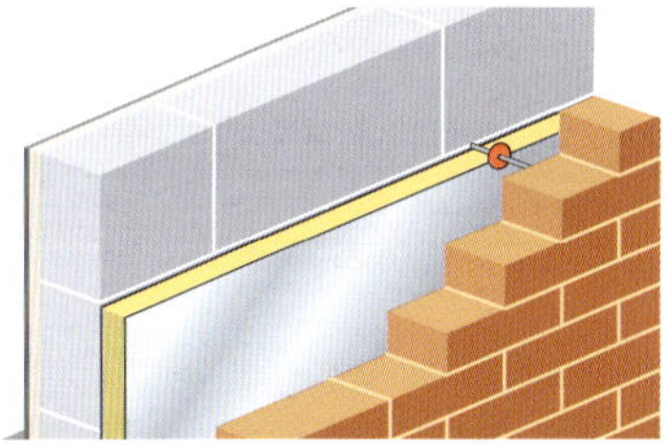 	The two walls that make up a cavity wall to comply with current building regulations. They are tied together with wall ties. *Regional variation: skin*
Lengthening joint  	Used to join two pieces of timber together to gain a longer length. Lengthening joints can be categorised as either structural or non-structural.
Lime	A fine powdered material traditionally used in mortars.
Lintel 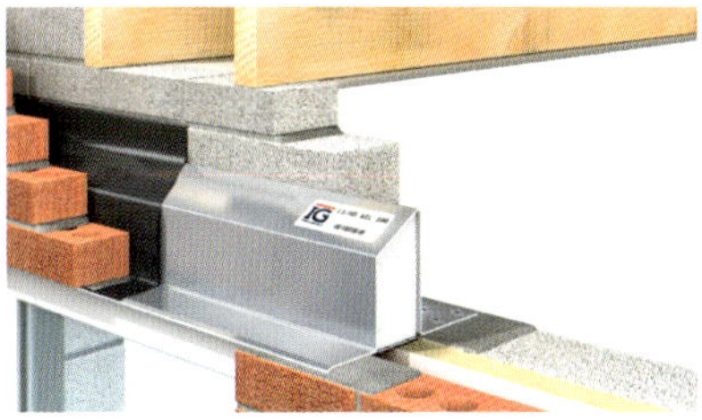 	A horizontal member for spanning an opening to support the structure above usually made from steel or concrete.

Industry term	Definition
Local exhaust ventilation (LEV)	An engineering control system to reduce exposure to airborne contaminants such as dust in the workplace. It vacuums away dust from the work area, and is usually attached to the tool being used. *Regional variation: extraction*
Mallet	A hammer with a large wooden head, used to knock pieces together or to drive dowels or chisels.
Manufactured boards	Manufactured boards are man-made boards and can include plywood, MDF (medium density fibre board), block board, chipboard, hard board and OSB (oriented strand board).
Marking gauge	A tool for scribing a line parallel to an edge, used in marking out.
Method statement	A description of the intended method of carrying out a task often linked to a risk assessment.
Misting spray	A bottle with a pressurising handle such as a kitchen cleaner spray bottle. Used to apply water to a diamond stone.
Mitre joint	An internal or external junction of a moulding, similar to a butt joint, but both pieces have been bevelled (usually at a 45° angle).

Industry term	Definition
Moisture content	The amount of moisture in a material, such as timber, expressed as a percentage.
Mortice and tenon joint	A very strong joint which is formed by a tongue-like piece or tenon. The tenon then fits into a mortice or slot cut into a second piece.
Mortice chisel	Mortice chisels are used for 'chopping out' joints (chiselling away the waste wood). They are particularly useful for cutting mortice joints as they are strong enough to withstand heavy blows with a mallet.
Mortice gauge	Similar to a marking gauge but it has two pins rather than one with a mechanism to allow the adjustment of the gap between the pins. Primarily used in marking out mortice and tenon or bridle joints.
Oil stone	A stone used for sharpening cutting tools, treated with oil to prevent steel particles from clogging the pores of the stone.
Orbital sander	An electric sander that moves the abrasive in an elliptical pattern. They are smaller machines than belt sanders and are used for finishing after the surface of the timber has been planed.
Ordnance bench mark (OBM)	They are a given height on an Ordnance Survey map. This fixed height is described as a value, eg so many metres above sea level (as calculated from the average sea height at Newlyn, Cornwall).

Industry term	Definition
Ovolo	A convex moulding having a cross-section in the form of a quarter of a circle or of an ellipse.
Pad saw	A thin saw blade in a tool pad used for cutting holes.
Panel saw	A fine-toothed saw with 8–10 teeth per 25mm, used for fine work.
Paring	The removal of small amounts of timber with the use of a chisel but without the use of a mallet.
Perimeter	The distance around an object or room.
Personal protective equipment (PPE)	This is defined in the Regulations as 'all equipment (including clothing affording protection against the weather) which is intended to be worn or held by a person at work and which protects against one or more risks to a person's health or safety.' For example, safety helmets, gloves, eye protection, high-visibility clothing, safety foot-wear and safety harnesses.
Philips (PH)	Screw with a cross slot in a simple star shape driven by a matching screwdriver.

Industry term	Definition
Photosynthesis	The process by which plants convert sunlight into food. Oxygen is a by-product.
Pitched roof	The most common roof, usually one with two slopes meeting at the central ridge.
Pitch marks	Small 'scoops' taken from the surface of timber when planing using rotary cutting tooling. The slower the feed speed, the better the finish on the timber and the smaller the pitch marks.
Planed all round (PAR)	Timber which has been planed on all four sides.
Planed square edge (PSE)	Timber which has been planed with a square face and edge.
Plough plane	Used for making grooves in timber.
Polycrystalline diamond (PCD)	Metal where industrial diamonds are bonded onto the surface making this very hard wearing.
Polyurethane (PU)	Commonly referred to as 'foam glue'. A yellow-brown resin that foams upon contact with air. It forms a strong water resistant bond and has excellent gap filling properties. This adhesive can be used to join damp timbers.

Industry term	Definition
Polyvinyl acetate (PVA)	Commonly referred to as 'white glue'. A resin dissolved in water. As the water evaporates the glue dries (goes off). Available in interior and exterior grades. It has good gap-filling properties and provides a strong permanent bond.
Portable appliance test (PAT)	PAT is a regular test carried out by a competent person (eg a qualified electrician) to ensure the tool is in a safe electrical condition. A sticker is placed on the tool after it has been tested.
Pozidriv (P2)	Screw with a cross slot shaped in a double star which must be driven by a matching screwdriver.
Pressure treatment	Wood impregnated with preserving and/or fire retarding chemicals under pressure in a vacuum tank.
Profiles	A shape cut into timber along its length. Examples would be a rebate or ovolo.
Prohibition notice	Issued by an HSE or local authority inspector when there is an immediate risk of personal injury. Receiving one means you are breaking health and safety regulations.
Proprietary	Manufactured and sold only by the owner of the patent, formula, brand name, or trademark associated with the product.
Purlin	A purlin is a strong large sectioned timber member which is fixed to the common rafters midway between the ridge and the wall plate and runs parallel to the wall and the ridge.
Push stick	A push stick is used for safety reasons when guiding wood being cut through a circular saw.
Quarter sawn	Term used in timber conversion. Cutting at a 90° angle from the growth rings on a log to produce a vertical and uniform pattern grain. The grain on the face of a quarter-sawn board will be parallel lines that are straight, tight and run the length of the board.

Industry term	Definition
Rebate	Rectangular groove or recess in the edge of a board which holds a panel or glass in a door or picture-frame.
Rebate plane	A plane for cutting rebates in timber. The plane has a fence to control the width of the rebate and a depth stop to set the depth of the rebate.
Recess	A space chopped out of timber to receive another piece of timber (joint) or piece of ironmongery, eg for a hinge.
Residual Current Device (RCD)	A device which acts as a failsafe trip system when using mains electricity.
Resin pocket	Resin pockets are formed in the growing tree as a result of damage. The pocket can contain liquid resin, which flows out readily when the pocket is sawn through.
Right angle	An angle that contains 90°.

Industry term	Definition
Ripping	To cut down the grain.
Rip saw	A saw with between 2 and 5 teeth per 25mm, used for cutting lengthways with the grain.
Risk assessment	An assessment of the hazards and risks associated with an activity and the reduction and monitoring of them.
Riving knife	Installed at the rear of the saw blade, this safety device reduces the risk of 'kick back' and accidental contact with the back of the saw blade.
Rod	Full-size drawing of the item being made. Using a working drawing as a guide, overall measurements are transferred to the rod. This will include the length and width of the frame, along with sizes of the materials to be used (*See also* Setting out).
Roofing joint	Joints used for a group of timber members fitted or joined together in a roof structure, to support the roof covering of a building.

Industry term	Definition
Router 	A versatile tool, it can be used to form grooves, rebates, housings, openings and profiles.
Sash cramp 	Cramp used for work, up to about 2m long. It is an adjustable steel bar with a bolt at one end and a fixed jaw at the other.
Scale 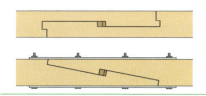 Scale: 1:1250	The ratio of the size on a drawing to the size of the real thing that it represents. It is impossible to fit a full-sized drawing of a building onto a sheet of paper, so it is necessary to scale the size of the building to enable it to fit. Scale rules are used to drawn scaled down buildings on paper.
Scarf joint 	A type of non-structural lengthening joint used for timber and veneers. The ends are put at a splay and glued together.
Services 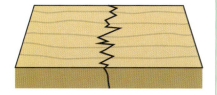	Those provided by the utility companies, eg gas, electricity and water.
Shakes	A type of timber defect. Shakes are cracks in the timber which appear due to excessive heat, frost or twisting due to wind during the growth of a tree. Depending upon the shape and the position, shakes can be classified as star shake, cup shake, ring shakes and heart shakes.

Industry term	Definition
Sliding bevel	A square which can be set to different angles to aid marking out. It is constructed of a hardwood stock, a sliding blade and a locking screw.
Slip stone	Small stone usually made from abrasive grit, used to hone and sharpen tools.
Sloping grain	If conversion is not parallel to the axis of tree, sloping grains occur which reduces timber's bending strength.
Slotted drive system (SDS)	Extremely powerful hammer drills used when extra power is required, eg for heavy duty jobs.
Smoothing plane	All-purpose plane used mainly for cleaning up and finishing work.
Solid walls	Walls of a thickness of one brick and greater. Unlike cavity walls, there are usually only two materials to consider; either bricks and mortar or blocks and mortar.
Specification	A contract document that gives information about the quality of materials and standards of workmanship required.

Industry term	Definition
Spelching	Damage at the end of a cut where the unsupported grain breaks away, ie when using a rip saw to cut across the grain.
Spokeshaves	Used to shape curved surfaces, consists of a blade fastened between two handles, blades come in straight, concave and convex curves.
Stair string	An inclined board each side of the stair to carry the treads (the top or horizontal surface of a step) and risers (the board that forms the face of the step).
Stress grading	A stress grade is defined as the classification of timber for structural purposes by means of either visual or machine grading. The grade is stamped on the timber which is either C16 (lower grade structural timber) or C24 (higher grade stuctural timber).
Stub mortice	One that does not pass through the full width of the timber.
Tang	The metal part of a chisel which fits into the handle.
Tangential conversion	Term used in timber conversion. Timber is sawn at a tangent to the heart. They are stronger when placed correctly, edge up with the load in the tangential axis.

Industry term	Definition
Tempering	Tempering is a process of heat treating, which is used to increase the toughness of steel.
Temporary bench mark (TBM)	Unlike an OBM, this is only temporary and is set up on site. These can be timber pegs surrounded with concrete.
Tender	Supplying a client with a fixed quotation for the work.
Tenon saw	Small saw used for the cutting of the shoulders of a tenon joint. A type of backed saw.
Through-and-through	Term used in timber conversion. Produces mostly tangentially sawn timber and some quarter sawn boards. Through-and-through timber is the most economical form of timber conversion.
Tolerance	Allowable variations between the specified measurement and the actual measurement.
Tongue and groove joint	A joint between two boards in which a raised area on the edge (tongue) of one board fits into a corresponding groove in the edge of the other to produce a flush surface. Commonly used in floorboards.

Industry term	Definition
Tooling	The part of the tool that cuts the timber, ie circular saw blade. This can be made from TCT, HSS or PCD metals.
Traverse	Move up and down or from side to side.
True	To be completely straight and square edged.
Try square	A square with a steel tongue in a wooden handle, used to mark or test a right angle.
Tungsten carbide tips (TCT)	Tungsten carbide tips on tooling are very abrasion resistant and can also withstand higher temperatures than standard high speed steel tools.
Twist	Spiral or corkscrew distortion in a longitudal direction of the board. *Regional variation: winding*
Twist drill	The most common and produced in largest quantity. It comprises a cutting point at the tip of a cylindrical shaft with helical flutes.
Unwrot	Timber which has been sawn.

Industry term	Definition
Urea-formaldehyde (UF)	An adhesive commonly referred to as powdered resin glue. It must be mixed with water before use and forms a strong water resistant bond when set.
Volatile organic compound (VOC)	The volatile organic compounds measure shows how much pollution a product will emit into the air when in use.
Waney edge	A type of timber defect. Boards or pieces of timber which, instead of being cut square, show the original curve of the log from which they are cut.
Water stone	Similar to an oil stone but made from natural stone. Removes less metal than man-made stones and can be lubricated with oil or water.
Wet rot	Decay of timber by fungi that attack wood that has high moisture content.
Wrot	Timber which has been planed.

Chapter 1
Unit 201: Health, safety and welfare in construction

A career in the building industry can be a very rewarding one, both personally and financially. However, building sites and workshops are potentially very dangerous places; there are many potential hazards in the construction industry. Many construction operatives (workers) are injured each year, some fatally. Regulations have been brought in over the years to reduce accidents and improve working conditions.

By reading this chapter you will know about:

1 The health and safety regulations, roles and responsibilities.
2 Accident and emergency reporting procedures and documentation.
3 Identifying hazards in the workplace.
4 Health and welfare in the workplace.
5 Handling materials and equipment safely.
6 Access equipment and working at heights.
7 Working with electrical equipment in the workplace.
8 Using personal protective equipment (PPE).
9 The cause of fire and fire emergency procedures.

HEALTH AND SAFETY LEGISLATION

According to the Health and Safety Executive (HSE) figures, in 2011/12:

- Forty-nine construction operatives were fatally injured. Twenty-three of these operatives were self-employed. This compares with an average of 59 fatalities over the previous five years, of which an average of 19 fatally injured construction operatives were self-employed.

- The rate of fatal injury per 100,000 construction operatives was 2.3, compared with a five-year average of 2.5.

- Construction industry operatives were involved in 28% of fatal injuries across all industry sectors and it accounts for the greatest number of fatal injuries in any industry sector.

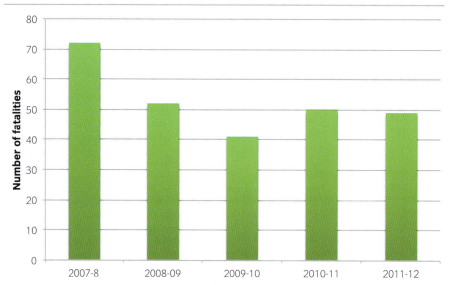

Number and rate of fatal injuries to workers in construction (RIDDOR)

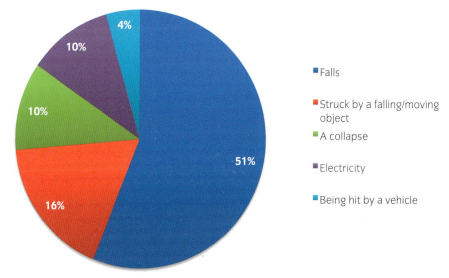

Proportion of fatalities in 2011/12 in construction

Health and safety legislation and great efforts made by the industry have made workplaces much safer in recent years. It is the responsibility of everyone involved in the building industry to continue to make it safer. Statistics are not just meaningless numbers – they represent injuries to real people. Many people believe that an accident will never happen to them, but it can. Accidents can:

- have a devastating effect on lives and families

- cost a lot financially in injury claims

- result in prosecution

- lead to job loss if an employee broke their company's safety policy.

Employers have an additional duty to ensure operatives have access to welfare facilities, eg drinking water, first aid and toilets, which will be discussed later in this chapter.

If everyone who works in the building industry pays close attention to health, safety and welfare, all operatives – including you – have every chance of enjoying a long, injury-free career.

UK HEALTH AND SAFETY REGULATIONS, ROLES AND RESPONSIBILITIES

Standard construction safety equipment

In the UK there are many laws (legislation) that have been put into place to make sure that those working on construction sites, and members of the public, are kept healthy and safe. If these laws and regulations are not obeyed then prosecutions can take place. Worse still, there is a greater risk of injury and damage to your health and the health of those around you.

The principle legislation which relates to health, safety and welfare in construction is:

- Health and Safety at Work Act (HASAWA) 1974

- Control of Substances Hazardous to Health (COSHH) Regulations 2002

- Reporting of Injuries, Diseases and Dangerous Occurrences Regulations (RIDDOR) 1995

- Construction, Design and Management (CDM) Regulations 2007

- Provision and Use of Work Equipment Regulations (PUWER) 1997

- Manual Handling Operations Regulations 1992

- Personal Protective Equipment (PPE) at Work Regulations 1992

- Work at Height Regulations 2005

- Lifting Operations and Lifting Equipment Regulations (LOLER) 1998

- Control of Noise at Work Regulations 2005

- Control of Vibration at Work Regulations 2005.

HEALTH AND SAFETY AT WORK ACT (HASAWA) 1974

The Health and Safety at Work Act (HASAWA) 1974 applies to all workplaces. Everyone who works on a building site or in a workshop is covered by this legislation. This includes employed and self-employed operatives, subcontractors, the employer and those delivering goods to the site. It not only protects those working, it also ensures the safety of anyone else who might be nearby.

KEY EMPLOYER RESPONSIBILITIES

The key employer health and safety responsibilities under HASAWA are to:

- provide a safe working environment

- provide safe access (entrance) and egress (exit) to the work area

- provide adequate staff training

- have a written health and safety policy in place

- provide health and safety information and display the appropriate signs

- carry out risk assessments

- provide safe machinery and equipment and to ensure it is well-maintained and in a safe condition

- provide adequate supervision to ensure safe practices are carried out

- involve trade union safety representatives, where appointed, in matters relating to health and safety

- provide personal protective equipment (**PPE**) free of charge, ensure the appropriate PPE is used whenever needed, and that operatives are properly supervised

- ensure materials and substances are transported, used and stored safely.

PPE

This is defined in the Personal Protective Equipment at Work Regulations 1992 as 'all equipment (including clothing affording protection against the weather) which is intended to be worn or held by a person at work and which protects against one or more risks to a person's health or safety.'

Risk assessments and method statements

The HASAWA requires that employers must carry out regular **risk assessments** to make sure that there are minimal dangers to their employees in a workplace.

Risk assessment

An assessment of the hazards and risks associated with an activity and the reduction and monitoring of them

Risk Assessment

Activity / Workplace assessed: Return to work after accident
Persons consulted / involved in risk assessment
Date:
Reviewed on:

Location:
Risk assessment reference number:
Review date:
Review by:

| Significant hazard | People at risk and what is the risk
Describe the harm that is likely to result from the hazard (e.g. cut, broken leg, chemical burn etc.) and who could be harmed (e.g. employees, contractors, visitors etc.) | Existing control measure
What is currently in place to control the risk? | Risk rating
Use matrix identified in guidance note
Likelihood (L)
Severity (S)
Multiply (L) * (S) to produce risk rating (RR) | | | | Further action required
What is required to bring the risk down to an acceptable level? Use hierarchy of control described in guidence note when considering the controls needed | Actioned to:
Who will complete the action? | Due date:
When will the action be complete by? | Completion date:
Initial and date once the action has been completed |
|---|---|---|---|---|---|---|---|---|---|---|
| | | | L | S | RR | L/M/H | | | | |
| Uneven floors | Operatives | Verbal warning and supervision | 2 | 1 | 2 | m | None applicable | Site supervisor | Active now | Ongoing |
| Steps | Operatives | Verbal warning | 2 | 1 | 2 | m | None applicable | Site supervisor | Active now | Ongoing |
| Staircases | Operatives | Verbal warning | 2 | 2 | 4 | m | None applicable | Site supervisor | Active now | Ongoing |

		Likelihood		
		1
Unlikely	2
Possible	3
Very likely
Severity	1
Slight/minor injuries/minor damage	1	2	3
	2
Medium injuries/significant damage	2	4	6
	3
Major injury/extensive damage	3	6	9

Likelihood
3 – Very likely
2 – possible
1 – Unlikely

Severity
3 – major injury/extensive damage
2 – medium injury/significant damage
1 – Slight/minor damage

1   – Low risk, action should be taken to reduce the risk if reasonably practicable
2, 3, 4 – Medium risk, is a significant risk and would require an appropriate level of resource
6 & 9 – High risk , may require considerable resource to mitigate. Control should focus on elimination of risk, if not possible control should be obtained by following the hierarchy of control

123 type risk assessment

A risk assessment is a legally-required tool used by employers to:

- identify work hazards

- assess the risk of harm arising from these hazards

- adequately control the risk.

Risk assessments are carried out as follows:

1 Identify the hazards. Consider the environment in which the job will be done. Which tools and materials will be used?

2 Identify who might be at risk. Think about operatives, visitors and members of the public.

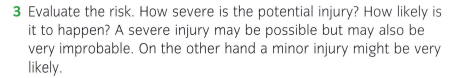

3 Evaluate the risk. How severe is the potential injury? How likely is it to happen? A severe injury may be possible but may also be very improbable. On the other hand a minor injury might be very likely.

4 If there is an unacceptable risk, can the job be changed? Could different tools or materials be used instead?

5 If the risk is acceptable, what measures can be taken to reduce the risk? This could be training, special equipment and using PPE.

6 Keep good records. Explain the findings of the risk assessment to the operatives involved. Update the risk assessment as required – there may be new machinery, materials or staff. Even adverse weather can bring additional risks.

A **method statement** is required by law and is a useful way of recording the hazards involved in a specific task. It is used to communicate the risk and precautions required to all those involved in the work. It should be clear, uncomplicated and easy to understand as it is for the benefit of those carrying out the work (and their immediate supervisors).

Inductions and tool box talks

Any new visitors to and operatives on a site will be given an induction. This will explain:

- the layout of the site

- any hazards of which they need to be aware

- the location of welfare facilities

- the assembly areas in case of emergency

- site rules.

Tool box talks are short talks given at regular intervals. They give timely safety reminders and outline any new hazards that may have arisen because construction sites change as they develop. Weather conditions such as extreme heat, wind or rain may create new hazards.

KEY EMPLOYEE RESPONSIBILITIES

The HASAWA covers the responsibilities of employees and subcontractors:

- You must work in a safe manner and take care at all times.

- You must make sure you do not put yourself or others at risk by your actions or inactions.

Method statement

A description of the intended method of carrying out a task, often linked to a risk assessment

INDUSTRY TIP

The Construction Skills Certification Scheme (CSCS) was set up in the mid-90s with the aim of improving site operatives' competence to reduce accidents and drive up on-site efficiency. Card holders must take a health and safety test. The colour of card depends on level of qualification held and job role. For more information see www.cscs.uk.com

ACTIVITY

Think back to your induction. Write down what was discussed. Did you understand everything? Do you need any further information? If you have not had an induction, write a list of the things you think you need to know.

INDUSTRY TIP

Remember, if you are unsure about any health and safety issue always seek help and advice.

- You must co-operate with your employer in regard to health and safety. If you do not you risk injury (to yourself or others), prosecution, a fine and loss of employment. Do not take part in practical jokes and horseplay.

- You must use any equipment and safeguards provided by your employer. For example, you must wear, look after and report any damage to the PPE that your employer provides.

- You must not interfere or tamper with any safety equipment.

- You must not misuse or interfere with anything that is provided for employees' safety.

FIRST AID AND FIRST-AID KITS

First aid should only be applied by someone trained in first aid. Even a minor injury could become infected and therefore should be cleaned and a dressing applied. If any cut or injury shows signs of infection, becomes inflamed or painful seek medical attention. An employer's first-aid needs should be assessed to indicate if a first-aider (someone trained in first aid) is necessary. The minimum requirement is to appoint a person to take charge of first-aid arrangements. The role of this appointed person includes looking after the first-aid equipment and facilities and calling the emergency services when required.

First-aid kits vary according to the size of the workforce. First-aid boxes should not contain tablets or medicines.

INDUSTRY TIP

The key employee health and safety responsibilities are to:
- work safely
- work in partnership with your employer
- report hazards and accidents as per company policy.

INDUSTRY TIP

Employees must not be charged for anything given to them or done for them by the employer in relation to safety.

INDUSTRY TIP

In the event of an accident, first aid will be carried out by a qualified first aider. First aid is designed to stabilise a patient for later treatment if required. The casualty may be taken to hospital or an ambulance may be called. In the event of an emergency you should raise the alarm.

ACTIVITY

Your place of work or training will have an appointed first-aider who deals with first aid. Find out who they are and how to make contact with them.

ACTIVITY

Find the first-aid kit in your workplace or place of training. What is inside it? Is there anything missing?

Eye wash
Foil blanket
Bandages
Cleaning wipes
Microporous tape
Safety pins
Scissors
Burn dressing
Resuscitation face shield
Nitrate gloves
Plasters

First-aid kit

SOURCES OF HEALTH AND SAFETY INFORMATION

Source	How they can help
Health and Safety Executive (HSE)	A government body which oversees health and safety in the workplace. It produces health and safety literature such as the **Approved Code of Practice** (ACoP).
Construction Skills	The construction industry training body produces literature and is directly involved with construction training.
The Royal Society for the Prevention of Accidents (ROSPA)	It produces literature and gives advice.
The Royal Society for Public Health	An independent, multi-disciplinary charity which is dedicated to the promotion and protection of collective human health and wellbeing.
Institution of Occupational Safety and Health (IOSH)	A chartered body for health and safety practitioners. The world's largest health and safety professional membership organisation.
The British Safety Council	It helps businesses with their health, safety and environmental management.

HEALTH AND SAFETY EXECUTIVE (HSE)

The HSE is a body set up by the government. The HSE ensures that the law is carried out correctly and has extensive powers to ensure that it can do its job. It can make spot checks in the workplace, bring the police, examine anything on the premises and take things away to be examined.

If the HSE finds a health and safety problem that breaks health and safety law it might issue an **improvement notice** giving the employer a set amount of time to correct the problem. For serious health and safety risks where there is a risk of immediate major injury, it can issue a **prohibition notice** which will stop all work on site until the health and safety issues are rectified. It may take an employer, employee, self-employed person (subcontractor) or anyone else

---

Approved Code of Practice

ACoP gives practical advice for those in the construction industry in relation to using machinery

INDUSTRY TIP

There are many other trade organisations, eg the Timber Research and Development Association (TRADA) which also offer advice on safe practices.

ACTIVITY

You have been asked to give a tool box talk because of several minor injuries involving tripping on site. What topics would you include in this talk?

INDUSTRY TIP

To find out more information on the sources in the table, enter their names into a search engine on the internet.

Improvement notice

Issued by an HSE or local authority inspector to formally notify a company that improvements are needed to the way it is working

Prohibition notice

Issued by an HSE or local authority inspector when there is an immediate risk of personal injury. They are not issued lightly and if you are on the receiving end of one, you are clearly breaking a health and safety regulation

involved with the building process to court for breaking health and safety legislation.

The HSE provides a lot of advice on safety and publishes numerous booklets and information sheets. One example of this is the Approved Code of Practice (ACoP) which applies to wood working machinery. The ACoP has a special legal status and employers and employees are expected to work within its guidelines.

The duties of the HSE are to:

- give advice

- issue improvement and prohibition notices

- caution

- prosecute

- investigate.

The Approved Code of Practice booklet is available free online

CONTROL OF SUBSTANCES HAZARDOUS TO HEALTH (COSHH) REGULATIONS 2002

The Control of Substances Hazardous to Health (COSHH) Regulations 2002 controls the use of dangerous substances, eg preservatives, fuel, solvents, adhesives, cement and oil-based paint. These have to be moved, stored and used safely without polluting the environment. It also covers hazardous substances produced while working, eg wood dust produced when sanding or drilling.

Hazardous substances may be discovered during the building process, eg lead-based paint or asbestos. These are covered by separate regulations.

When considering substances and materials that may be hazardous to health an employer should do the following to comply with COSHH:

- Read and check the COSHH safety data sheet that comes with the product. It will outline any hazards associated with the product and the safety measures to be taken.

- Check with the supplier if there are any known risks to health.

- Use the trade press to find out if there is any information about this substance or material.

- Use the HSE website, or other websites, to check any known issues with the substance or material.

When assessing the risk of a potentially dangerous substance or material it is important to consider how operatives could be exposed to it. For example:

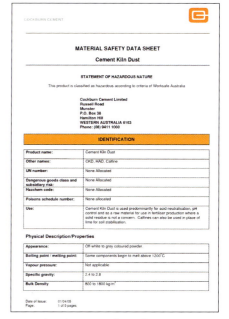

Example of COSHH data sheet

- by breathing in gas or mist

- by swallowing it,

- by getting into their eyes

- through their skin, either by contact or through cuts.

Safety data sheets

Products you use may be 'dangerous for supply'. If so, they will have a label that has one or more hazard symbols. Some examples are given here.

These products include common substances in everyday use such as paint, bleach, solvent or fillers. When a product is 'dangerous for supply', by law, the supplier must provide you with a safety data sheet. Note: medicines, pesticides and cosmetic products have different legislation and don't have a safety data sheet. Ask the supplier how the product can be used safely.

Safety data sheets can be hard to understand, with little information on measures for control. However, to find out about health risks and emergency situations, concentrate on:

- Sections 2 and 16 of the sheet, which tell you what the dangers are;
- Sections 4-8, which tell you about emergencies, storage and handling.

Since 2009, new international symbols have been gradually replacing the European symbols. Some of them are similar to the European symbols, but there is no single word describing the hazard. Read the hazard statement on the packaging and the safety data sheet from the supplier.

European symbols

Toxic   Very toxic   Harmful   Irritant

Highly flammable   Extremely flammable   Explosive   Dangerous to the environment

Oxidising   Corrosive

New International symbols

Hazard checklist

- ☐ Does any product you use have a danger label?
- ☐ Does your process produce gas, fume, dust, mist or vapour?
- ☐ Is the substance harmful to breathe in?
- ☐ Can the substance harm your skin?
- ☐ Is it likely that harm could arise because of the way you use or produce it?
- ☐ What are you going to do about it?
  - Use something else?
  - Use it in another, safer way?
  - Control it to stop harm being caused?

CONTROL MEASURES

The control measures below are in order of importance.

1 Eliminate the use of the harmful substance and use a safer one. For instance, swap high **VOC** oil-based paint for a lower VOC water-based paint.

2 Use a safer form of the product. Is the product available ready-mixed? Is there a lower strength option that will still do the job?

VOC
The volatile organic compounds measure shows how much pollution a product will emit into the air when in use

INDUSTRY TIP
Product data sheets are free and have to be produced by the supplier of the product.

3 Change the work method to emit less of the substance. For instance, applying paint with a brush releases fewer VOCs into the air than spraying paint. Wet grinding produces less dust than dry grinding.

4 Enclose the work area so that the substance does not escape. This can mean setting up a tented area or closing doors.

5 Use extraction or filtration (eg a dust bag) in the work area.

6 Keep operatives in the area to a minimum.

7 Employers must provide appropriate PPE.

Paint with high VOC content

European symbols

Toxic   Very toxic   Harmful   Irritant

Highly flammable   Extremely flammable   Explosive   Dangerous to the environment

Oxidising   Corrosive

New International symbols

Toxic   May explode when heated   Irritant

Causes fire   Explosive   Dangerous to the environment

Intensifies fire   Long term health hazard   Corrosive

COSHH symbols. The international symbols will replace the European symbols in 2015.

REPORTING OF INJURIES, DISEASES AND DANGEROUS OCCURRENCES REGULATIONS (RIDDOR) 1995

Despite all the efforts put into health and safety, incidents still happen. The Reporting of Injuries, Diseases and Dangerous Occurrences Regulations (RIDDOR) 1995 state that employers must report to the HSE all accidents that result in an employee needing more than seven days off work. Diseases and dangerous occurrences must also be reported. A serious occurrence which has not caused an injury (a near miss) should still be reported because next time it happens things might not work out as well.

Below are some examples of injuries, diseases and dangerous occurrences which would need to be reported:

- A joiner cuts off a finger while using a circular saw.

- A plumber takes a week off after a splinter in her hand becomes infected.

- A ground operative contracts **leptospirosis**.

- A labourer contracts dermatitis (a serious skin problem) after contact with an irritant substance.

- A scaffold suffers a collapse following severe weather, unauthorised alteration or overloading but no-one is injured.

Leptospirosis

Also known as Weil's disease, this is a serious disease spread by rats and cattle

The purpose of RIDDOR is to enable the HSE to investigate serious incidents and collate statistical data. This information is used to help reduce the number of similar accidents happening in future and to make the workplace safer.

An F2508 injury report form

Although minor accidents and injuries are not reported to HSE, records must be kept. Accidents must be recorded in the accident book. This provides a record of what happened and is useful for future reference. Trends may become apparent and the employer may take action to try and prevent that particular type of accident occurring again.

CONSTRUCTION, DESIGN AND MANAGEMENT (CDM) REGULATIONS 2007

The Construction, Design and Management (CDM) Regulations 2007 focus attention on the effective planning and management of construction projects, from the design concept through to maintenance and repair. The aim is for health and safety considerations to be integrated into a project's development, rather than be an inconvenient afterthought. The CDM Regulations reduce the risk of harm to those that have to work on or use the structure throughout its life, from construction through to **demolition**.

The CDM regulations play a role in safety during demolition

Demolition

When something, often a building, is completely torn down and destroyed

CDM Regulations protect workers from the construction to demolition of large and complex structures

The CDM Regulations apply to all projects except for those arranged by private clients, ie work that isn't in furtherance of a business interest. Property developers need to follow the CDM Regulations.

Under the CDM Regulations, the HSE must be notified where the construction work will take:

- more than 30 working days or

- 500 working days in total, ie if 100 people work for 5 days (500 working days) the HSE will have to be notified.

DUTY HOLDERS

Under the CDM Regulations there are several duty holders, each with a specific role.

Duty holder	Role
Client	This is the person or organisation who wishes to have the work done. The client will check that: ■ all the team members are competent ■ the management is suitable ■ sufficient time is allowed for all stages of the project ■ welfare facilities are in place before construction starts. HSE notifiable projects require that the client appoints a CDM co-ordinator and principal contractor, and provides access to a health and safety file.
CDM co-ordinator	Appointed by the client, the co-ordinator advises and assists the client with CDM duties. The co-ordinator notifies the HSE before work starts. This role involves the co-ordination of the health and safety aspects of the design of the building and ensures good communication between the client, designers and contractors.
Designer	At the design stages the designer removes hazards and reduces risks. The designer provides information about the risks that cannot be eliminated. Notifiable projects require that the designer checks that the client is aware of their CDM duties and that a CDM co-ordinator has been appointed. The designer will also supply information for the health and safety file.
Principal contractor	The principal contractor will plan, manage and monitor the construction in liaison with any other involved contractors. This involves developing a written plan and site rules before the construction begins. The principle contractor ensures that the site is made secure and suitable welfare facilities are provided from the start and maintained throughout construction. The principal contractor will also make sure that all operatives have site inductions and any further training that might be required to make sure the workforce is competent.
Contractor	Subcontractors and self-employed operatives will plan, manage and monitor their own work and employees, co-operating with any main contractor in relation to site rules. Contractors will make sure that all operatives have any further training that might be required to make sure they are competent. A contractor also reports any incidents under RIDDOR to the principal contractor.
Operatives	Operatives need to check their own competence: Can you carry out the task you have been asked to do safely? Have you been trained to do this type of activity? Do you have the correct equipment to carry out this activity? You must follow all the site health and safety rules and procedures and fully co-operate with the rest of the team to ensure the health and safety of other operatives and others who may be affected by the work. Any health and safety issues must be reported.

A client, a contractor and an operative looking over building plans ahead of construction

ACTIVITY

What would you do if you spotted any of these hazards?

WELFARE FACILITIES REQUIRED ON SITE UNDER THE CDM REGULATIONS

The table below shows the welfare facilities that must be available on site.

Facility	Site requirement
Sanitary conveniences (toilets) 	■ Suitable and sufficient toilets should be provided or made available. ■ Toilets should be adequately ventilated and lit and should be clean. ■ Separate toilet facilities should be provided for men and women.
Washing facilities 	■ Sufficient facilities must be available, and include showers if required by the nature of the work. ■ They should be in the same place as the toilets and near any changing rooms. ■ There must be a supply of clean hot (or warm) and cold running water, soap and towels. ■ There must be separate washing facilities provided for men and women unless the area is for washing hands and the face only.

Facility	Site requirement
Clean drinking water 	■ This must be provided or made available. ■ It should be clearly marked by an appropriate sign. ■ Cups should be provided unless the supply of drinking water is from a water fountain.
Changing rooms and lockers 	■ Changing rooms must be provided or made available if operatives have to wear special clothing and if they cannot be expected to change elsewhere. ■ There must be separate rooms for, or separate use of rooms by, men and women where necessary. ■ The rooms must be have seating and include, where necessary, facilities to enable operatives to dry their special clothing and their own clothing and personal effects. ■ Lockers should also be provided.
Rest rooms or rest areas 	■ They should have enough tables and seating with backs for the number of operatives likely to use them at any one time. ■ Where necessary, rest rooms should include suitable facilities for pregnant women or nursing mothers to rest lying down. ■ Arrangements must be made to ensure that meals can be prepared, heated and eaten. It must also be possible to boil water.

ACTIVITY

What facilities are provided at your workplace or place of training?

PROVISION AND USE OF WORK EQUIPMENT REGULATIONS (PUWER) 1997

The Provision and Use of Work Equipment Regulations (PUWER) 1997 place duties on:

■ people and companies who own, operate or have control over work equipment

■ employers whose employees use work equipment.

Work equipment can be defined as any machinery, appliance, apparatus, tool or installation for use at work (whether exclusively or not). This includes equipment which employees provide for their own use at work. The scope of work equipment is therefore extremely wide. The use of work equipment is also very widely interpreted and, according to the HSE, means 'any activity involving work equipment and includes starting, stopping, programming, setting, transporting, repairing, modifying, maintaining, servicing and cleaning.' It includes equipment such as diggers, electric planers, stepladders, hammers or wheelbarrows.

Under PUWER, work equipment must be:

- suitable for the intended use

- safe to use

- well maintained

- inspected regularly.

Regular inspection is important as a tool that was safe when it was new may no longer be safe after considerable use.

Additionally, work equipment must only be used by people who have received adequate instruction and training. Information regarding the use of the equipment must be given to the operator and must only be used for what it was designed to do.

Protective devices, eg emergency stops, must be used. Brakes must be fitted where appropriate to slow down moving parts to bring the equipment to a safe condition when turned off or stopped. Equipment must have adequate means of isolation. Warnings, either by signs or other means such as sounds or lights, must be used as appropriate. Access to dangerous parts of the machinery must be controlled. Some work equipment is subject to additional health and safety legislation which must also be followed.

Employers who use work equipment must manage the risks. ACoPs (see page 9) have been developed in line with PUWER. The ACoPs have a special legal status, as outlined in the introduction to the PUWER ACoP:

> *Following the guidance is not compulsory and you are free to take other action. But if you do follow the guidance you will normally be doing enough to comply with the law. Health and safety inspectors seek to secure compliance with the law and may refer to this guidance as illustrating good practice.*

MANUAL HANDLING OPERATIONS REGULATIONS 1992

Employers must try and avoid manual handling within reason if there is a possibility of injury. If manual handling cannot be avoided then they must reduce the risk of injury by means of a risk assessment.

INDUSTRY TIP

Abrasive wheels are used for grinding. Under PUWER these wheels can only be changed by someone who has received training to do this. Wrongly fitted wheels can explode!

ACTIVITY

All the tools you use for your work are covered by PUWER. They must be well maintained and suitable for the task. A damaged head on a bolster chisel must be reshaped. A split shaft on a joiner's wood chisel must be repaired. Why would these tools be dangerous in a damaged condition? List the reasons.

An operative lifting heavy bricks

LIFTING AND HANDLING

Incorrect lifting and handling is a serious risk to your health. It is very easy to injure your back – just ask any experienced builder. An injured back can be very unpleasant, so it's best to look after it.

Here are a few things to consider when lifting:

- Assess the load. Is it too heavy? Do you need assistance or additional training? Is it an awkward shape?

- Can a lifting aid be used, such as any of the below?

Wheelbarrow

Gin lift

Scissor lift

Kerb lifter

- Does the lift involve twisting or reaching?

- Where is the load going to end up? Is there a clear path? Is the place it's going to be taken to cleared and ready?

How to lift and place an item correctly

If you cannot use a machine, it is important that you keep the correct posture when lifting any load. The correct technique to do this is known as **kinetic lifting**. Always lift with your back straight, elbows in, knees bent and your feet slightly apart.

Kinetic lifting

A method of lifting that ensures that the risk of injury is reduced

Safe kinetic lifting technique

ACTIVITY

Try it out. Place a box on the floor and lift it using the technique shown.

ACTIVITY

Consider this list of materials: plywood, cement, aggregates, sawn timber joists, glass, drainage pipes, and kerbs. Make a table to show how you would transport and stack them around your place of work.

INDUSTRY TIP

Most workplace injuries are a result of manual handling. Remember pushing or pulling an object still comes under the Manual Handling Operations Regulations.

When placing the item, again be sure to use your knees and beware of trapping your fingers. If stacking materials, be sure that they are on a sound level base and on bearers if required.

Heavy objects that cannot easily be lifted by mechanical methods can be lifted by several people. It is important that one person in the team is in charge, and that lifting is done in a co-operative way. It has been known for one person to fall down and the others then drop the item!

CONTROL OF NOISE AT WORK REGULATIONS 2005

Under the Control of Noise at Work Regulations 2005, duties are placed on employers and employees to reduce the risk of hearing damage to the lowest reasonable level practicable. Hearing loss caused by work is preventable. Hearing damage is permanent and cannot be restored once lost.

ACTIVITY

Watch this link to find out more about hearing loss and damage: www.hse.gov.uk/noise/video/hearingvideo.htm

EMPLOYER'S DUTIES UNDER THE REGULATIONS

An employer's duties are:

- To carry out a risk assessment and identify who is at risk.

- To eliminate or control its employees exposure to noise at the workplace and to reduce the noise as far as practicable.

- To provide suitable hearing protection.

- To provide health surveillance to those identified as at risk by the risk assessment.

- To provide information and training about the risks to their employees as identified by the risk assessment.

EMPLOYEES' DUTIES UNDER THE REGULATIONS

Employees must:

- Make full and proper use of personal hearing protectors provided to them by their employer.

- If they discover any defect in any personal hearing protectors or other control measures they must report it to their employer as soon as is practicable.

Ear defenders

Ear plugs

NOISE LEVELS

Under the Regulations, specific actions are triggered at specific noise levels. Noise is measured in decibels and shown as dB (a). The two main action levels are 80dB (a) and 85dB (a).

Requirements at 80dB (a) to 85dB (a):

- Assess the risk to operatives' health and provide them with information and training.

- Provide suitable ear protection free of charge to those who request ear protection.

Requirements above 85dB (a):

- Reduce noise exposure as far as practicable by means other than ear protection.

- Set up an ear protection zone using suitable signage and segregation.

- Provide suitable ear protection free of charge to those affected and ensure they are worn.

PERSONAL PROTECTIVE EQUIPMENT (PPE) AT WORK REGULATIONS 1992

Employees and subcontractors must work in a safe manner. Not only must they wear the PPE that their employers provide they must also look after it and report any damage to it. Importantly, employees must not be charged for anything given to them or done for them by the employer in relation to safety.

The hearing and respiratory PPE provided for most work situations is not covered by these Regulations because other regulations apply to it. However, these items need to be compatible with any other PPE provided.

The main requirement of the Regulations is that PPE must be supplied and used at work wherever there are risks to health and safety that cannot be adequately controlled in other ways.

The Regulations also require that PPE is:

- included in the method statement

- properly assessed before use to ensure it is suitable

- maintained and stored properly

- provided to employees with instructions on how they can use it safely

- used correctly by employees.

An employer cannot ask for money from an employee for PPE, whether it is returnable or not. This includes agency workers if they are legally regarded as employees. If employment has been terminated and the employee keeps the PPE without the employer's permission, then, as long as it has been made clear in the contract of employment, the employer may be able to deduct the cost of the replacement from any wages owed.

Using PPE is a very important part of staying safe. For it to do its job properly it must be kept in good condition and used correctly. If any damage does occur to an article of PPE it is important that this is reported and it is replaced. It must also be remembered that PPE is a last line of defence and should not be used in place of a good safety policy!

ACTIVITY

Check the date on your safety helmet. Always update your safety helmet if it is out of date.

INDUSTRY TIP

Remember, you also have a duty of care for your own health.

A site safety sign showing the PPE required to work there

The following table shows the type of PPE used in the workplace and explains why it is important to store, maintain and use PPE correctly. It also shows why it is important to check and report damage to PPE.

PPE	Correct use
Hard hat/safety helmet 	Hard hats must be worn when there is danger of hitting your head or danger of falling objects. They often prevent a wide variety of head injuries. Most sites insist on hard hats being worn. They must be adjusted to fit your head correctly and must not be worn back to front! Check the date of manufacture as plastic can become brittle over time. Solvents, pens and paints can damage the plastic too.
Toe-cap boots or shoes  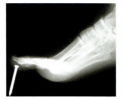 Safety boots A nail in a construction worker's foot.	Toe-cap boots or shoes are worn on most sites as a matter of course and protect the feet from heavy falling objects. Some safety footwear has additional insole protection to help prevent nails going up through the foot. Toe caps can be made of steel or lighter plastic.
Ear defenders and plugs   Ear defenders Ear plugs	Your ears can be very easily damaged by loud noise. Ear protection will help prevent hearing loss while using loud tools or if there is a lot of noise going on around you. When using earplugs always ensure your hands are clean before handling the plugs as this reduces the risk of infection. If your ear defenders are damaged or fail to make a good seal around your ears have them replaced.
High visibility (hi-viz) jacket 	This makes it much easier for other people to see you. This is especially important when there is plant or vehicles moving in the vicinity.
Goggles and safety glasses   Safety goggles Safety glasses	These protect the eyes from dust and flying debris while you are working. It has been known for casualties to be taken to hospital after dust has blown up from a dry mud road. You only get one pair of eyes, look after them!

PPE	Correct use
Dust masks and respirators  Dust mask　　　Respirator	Dust is produced during most construction work and it can be hazardous to your lungs. It can cause all sorts of ailments from asthma through to cancer. Wear a dust mask to filter this dust out. You must ensure it is well fitted. Another hazard is dangerous gases such as solvents. A respirator will filter out hazardous gases but a dust mask will not! Respirators are rated P1, P2 and P3, with P3 giving the highest protection.
Gloves Latex glove　　　Nitrile glove 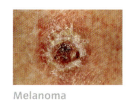 Gauntlet gloves　　Leather gloves	Gloves protect your hands. Hazards include cuts, abrasions, dermatitis, chemical burns or splinters. Latex and nitrile gloves are good for fine work, although some people are allergic to latex. Gauntlets provide protection from strong chemicals. Other types of gloves provide good grip and protect the fingers. 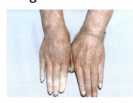 A chemical burn as a result of not wearing safety gloves
Sunscreen Suncream　　　Melanoma	Another risk, especially in the summer months, is sunburn. Although a good tan is sometimes considered desirable, over-exposure to the sun can cause skin cancer such as melanoma. When out in the sun, cover up and use sunscreen (ie suncream) on exposed areas of your body to prevent burning.
Preventing HAVS	Vibration white finger (VWS) is a symptom of an industrial injury known as HAVS and is caused by using vibrating power tools (such as a hammer drill, vibrating poker and vibrating plate) for a long time. This injury is controlled by limiting the time such power tools are used. For more information see page 31.

For more information see page 31.

ACTIVITY

You are working on a site and a brick falls on your head. Luckily, you are doing as you have been instructed and you are wearing a helmet. You notice that the helmet has a small crack in it. What do you do?

1 Carry on using it as your employer will charge you for a new one, after all it is only a small crack.
2 Take it to your supervisor as it will no longer offer you full protection and it will need replacing.
3 Buy a new helmet because the old one no longer looks very nice.

INDUSTRY TIP

The most important pieces of PPE when using a disc cutter are dust masks, glasses and ear protection.

WORK AT HEIGHT REGULATIONS 2005

The Work at Height Regulations 2005 put several duties upon employers:

- Working at height should be avoided if possible.

- If working at height cannot be avoided, the work must be properly organised with risk assessments carried out.

- Risk assessments should be regularly updated.

- Those working at height must be trained and competent.

- A method statement must be provided.

Operatives working at height as a roof is lifted into place

Several points should be considered when working at height:

- How long is the job expected to take?

- What type of work will it be? It could be anything from fitting a single light bulb, through to removing a chimney or installing a roof.
  - How is the access platform going to be reached? By how many people?
  - Will people be able to get on and off the structure safely? Could there be overcrowding?

- What are the risks to passers-by? Could debris or dust blow off and injure anyone on the road below?

- What are the conditions like? Extreme weather, unstable buildings and poor ground conditions need to be taken into account.

A cherry picker can assist you when working at height

ACCESS EQUIPMENT AND SAFE METHODS OF USE

The means of access should only be chosen after a risk assessment has been carried out. There are various types of access.

Ladders

Ladders are normally used for access onto an access platform. They are not designed for working from except for light, short-duration work. A ladder should lean at an angle of 75°, ie one unit out for every four units up.

Strong upper resting point

Adequate lap on extension ladders

Ground back slope not exceeding 6°

Ground side slope not exceeding 16°, clean and free of slippery algae and moss

Using a ladder correctly

Roof ladder

Resting ladders on plastic guttering can cause it to bend and break

The following images show how to use a ladder or stepladder safely.

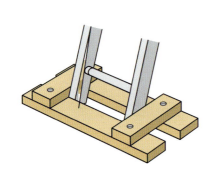

A ladder secured at the base.

A ladder secured at the top of a platform for working from.

Access ladders should extend 1m above the landing point to provide a strong handhold.

Certain stepladders are unsafe to work from the top three rungs.

Don't overreach, and stay on the same rung.

Grip the ladder when climbing and remember to keep three points of contact.

Stepladders

Stepladders are designed for light, short-term work.

Working from the side can make stepladders unstable. Do not overreach

Don't stand on the top three steps

Stepladder is fully open

Locked open firm and level on the ground

Using a stepladder correctly

Trestles

This is a working platform used for work of a slightly longer duration.

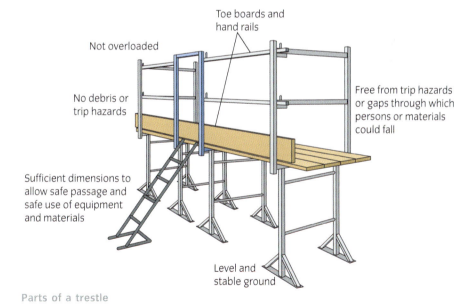

Not overloaded

Toe boards and hand rails

No debris or trip hazards

Free from trip hazards or gaps through which persons or materials could fall

Sufficient dimensions to allow safe passage and safe use of equipment and materials

Level and stable ground

Parts of a trestle

Tower scaffold

These are usually proprietary (manufactured) and are made from galvanised steel or lightweight aluminium alloy. They must be erected by someone competent in the erection and dismantling of mobile scaffolds.

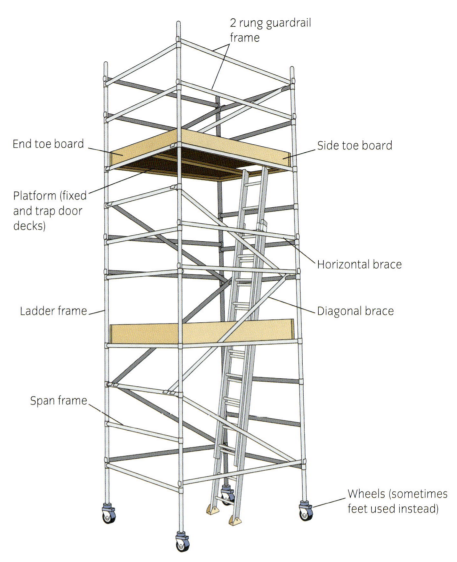

2 rung guardrail frame

End toe board

Side toe board

Platform (fixed and trap door decks)

Horizontal brace

Ladder frame

Diagonal brace

Span frame

Wheels (sometimes feet used instead)

Parts of a tower scaffold

To use a tower scaffold safely:

- Always read and follow the manufacturer's instruction manual.

- Only use the equipment for what it is designed for.

- The wheels or feet of the tower must be in contact with a firm surface.

- Outriggers should be used to increase stability. The maximum height given in the manufacturer's instructions must not be exceeded.

- The platform must not be overloaded.

- The platform should be unloaded (and reduced in height if required) before it is moved.

- Never move a platform, even a small distance, if it is occupied.

Tubular scaffold

This comes in two types:

- independent scaffold has two sets of standards or uprights

- putlog scaffold is built into the brickwork.

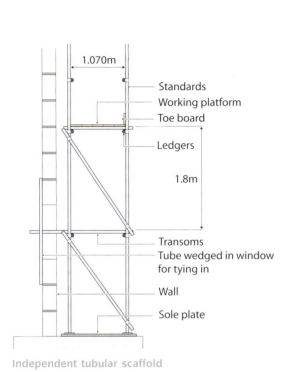

Independent tubular scaffold

Putlog tubular scaffold

Tubular scaffold is erected by specialist scaffolding companies and often requires structural calculations. Only trained and competent scaffold erectors should alter scaffolding. Access to a scaffold is usually via a tied ladder with three rungs projecting above the step off at platform level.

OUR HOUSE

You have been asked to complete a job that requires gaining access to the roof level of a two-storey building. What equipment would you choose to get access to the work area? What things would you take into consideration when choosing the equipment? Take a look at 'Our House' as a guide for working on a two-storey building.

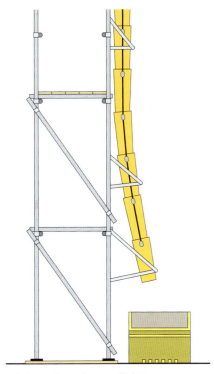

A debris chute for scaffolding

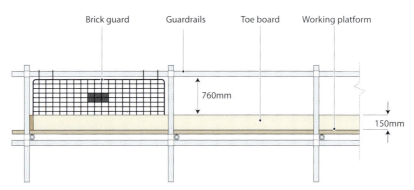

Brick guard   Guardrails   Toe board   Working platform

760mm

150mm

A safe working platform on a tubular scaffold

All scaffolding must:

- not have any gaps in the handrail or toe boards

- have a safe system for lifting any materials up to the working height

- have a safe system of debris removal.

Fall protection devices include:

- harnesses and lanyards

- safety netting

- air bags.

A harness and lanyard or safety netting will stop a person falling too far, leaving them suspended in the air. Air bags (commonly known as 'bouncy castles') are set up on the ground and inflated. If a person falls, they will have a soft landing. Air bags have fallen out of favour somewhat as some operatives use them as an easy way to get off the working platform – not the purpose they were intended for!

A safe scaffolding set up

LIFTING OPERATIONS AND LIFTING EQUIPMENT REGULATIONS (LOLER) 1998

The Lifting Operations and Lifting Equipment Regulations (LOLER) 1998 put responsibility upon employers to ensure that the lifting equipment provided for use at work is:

- strong and stable enough for the particular use and marked to indicate safe working loads

- positioned and installed to minimise any risks

- used safely, ie the work is planned, organised and performed by competent people

- subject to on-going thorough examination and, where appropriate, inspection by competent people.

THE CONTROL OF VIBRATION AT WORK REGULATIONS 2005

Vibration white finger or hand-arm vibration syndrome (HAVS), see page 23, is caused by using vibrating tools such as hammer drills, vibrating pokers or hand held breakers over a long period of time. The most efficient and effective way of controlling exposure to hand-arm vibration is to look for new or alternative work methods which remove or reduce exposure to vibration.

Follow these steps to reduce the effects of HAVS:

- Always use the right tool for each job.

- Check tools before using them to make sure they have been properly maintained and repaired to avoid increased vibration caused by faults or general wear.

- Make sure cutting tools are kept sharp so that they remain efficient.

- Reduce the amount of time you use a tool in one go, by doing other jobs in between.

- Avoid gripping or forcing a tool or work piece more than you have to.

- Encourage good blood circulation by:
  - keeping warm and dry (when necessary, wear gloves, a hat, waterproofs and use heating pads if available)
  - giving up or cutting down on smoking because smoking reduces blood flow
  - massaging and exercising your fingers during work breaks.

Damage from HAVS can include the inability to do fine work and cold can trigger painful finger blanching attacks (when the ends of your fingers go white).

An operative taking a rest from using a power tool

Don't use power tools for longer than you need to

CONSTRUCTION SITE HAZARDS

DANGERS ON CONSTRUCTION SITES

Study the drawing of a building site. There is some demolition taking place, as well as construction. How many hazards can you find? Discuss your answers.

Dangers	Discussion points
Head protection	The operatives are not wearing safety helmets, which would prevent them from hitting their head or from falling objects.
Poor housekeeping	The site is very untidy. This can result in slips trips and falls and can pollute the environment. An untidy site gives a poor company image. Offcuts and debris should be regularly removed and disposed of according to site policy and recycled if possible.
Fire	There is a fire near a building; this is hazardous. Fires can easily become uncontrollable and spread. There is a risk to the structure and, more importantly, a risk of operatives being burned. Fires can also pollute to the environment.

Dangers	Discussion points
Trip hazards	Notice the tools and debris on the floor. The scaffold has been poorly constructed. There is a trip hazard where the scaffold boards overlap.
Chemical spills	There is a drum leaking onto the ground. This should be stored properly – upright and in a lockable metal shed or cupboard. The leak poses a risk of pollution and of chemical burns to operatives.
Falls from height	The scaffold has handrails missing. The trestle working platform has not been fitted with guard rails. None of the operatives are wearing hard hats for protection either.
Noise	An operative is using noisy machinery with other people nearby. The operative should be wearing ear PPE, as should those working nearby. Better still, they should be working elsewhere if at all possible, isolating themselves from the noise.
Electrical	Some of the wiring is 240V as there is no transformer, it's in poor repair and it's also dragging through liquid. This not only increases the risk of electrocution but is also a trip hazard.
Asbestos or other hazardous substances	Some old buildings contain **asbestos** roofing which can become a hazard when being demolished or removed. Other potential hazards include lead paint or mould spores. If a potentially hazardous material is discovered a supervisor must be notified immediately and work must stop until the hazard is dealt with appropriately.

Asbestos

A naturally occurring mineral that was commonly used for a variety of purposes including: **insulation**, fire protection, roofing and guttering. It is extremely hazardous and can cause a serious lung disease known as asbestosis

Insulation

A material that reduces or prevents the transmission of heat

Cables can be a trip hazard on site

Boiler suit

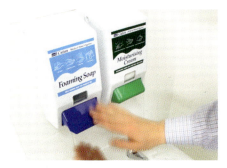

Hand cleaner

PERSONAL HYGIENE

Working in the construction industry can be very physical, and it's likely to be quite dirty at times. Therefore you should take good care with your personal hygiene. This involves washing well after work. If contaminants are present, then wearing a protective suit, such as a boiler suit, that you can take off before you go home will prevent contaminants being taken home with you.

You should also wash your hands after going to the toilet and before eating. This makes it safer to eat and more pleasant for others around you. The following step by steps show a safe and hygienic way to wash your hands.

STEP 1 Apply soap to hands from the dispenser.

STEP 2 Rub the soap into the lather and cover your hands with it, including between your fingers.

STEP 3 Rinse hands under a running tap removing all of the soap from your hands.

STEP 4 Dry your hands using disposable towels. Put the towel in the bin once your hands are dry.

WORKING WITH ELECTRICITY

Electricity is a very useful energy resource but it can be very dangerous. Electricity must be handled with care! Only trained, competent people can work with electrical equipment.

THE DANGERS OF USING ELECTRICAL EQUIPMENT

The main dangers of electricity are:

- shock and burns (a 230V shock can kill)

- electrical faults which could cause a fire

- an explosion where an electrical spark has ignited a flammable gas.

VOLTAGES

Generally speaking, the lower the voltage the safer it is. However, a low voltage is not necessarily suitable for some machines, so higher voltages can be found. On site, 110V (volts) is recommended and this is the voltage rating most commonly used in the building industry. This is converted from 230V by use of a transformer.

230V (commonly called 240V) domestic voltage is used on site as battery chargers usually require this voltage. Although 230V is often used in workshops, 110V is recommended.

410V (otherwise known as 3 phase) is used for large machinery, such as joinery shop equipment.

Voltages are nominal, ie they can vary slightly.

110V 1 phase – yellow

230V 1 phase – blue

BATTERY POWER

Battery power is much safer than mains power. Many power tools are now available in battery-powered versions. They are available in a wide variety of voltages from 3.6V for a small screwdriver all the way up to 36V for large masonry drills.

The following images are all examples of battery powered tools you may come across in your workplace or place of training.

410V 3 phase – red

Battery drill   Battery-powered planer   Battery-powered jigsaw

WIRING

The wires inside a cable are made from copper, which conducts electricity. The copper is surrounded by a plastic coating that is colour coded. The three wires in a cable are the live (brown), which works with the neutral (blue) to conduct electricity, making the appliance work. The earth (green and yellow stripes) prevents electrocution if the appliance is faulty or damaged.

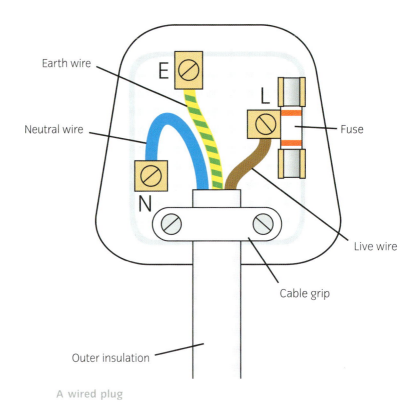

A wired plug

POWER TOOLS AND CHECKS

Power tools should always be checked before use. Always inform your supervisor if you find a fault. The tool will need to be repaired, and the tool needs to be kept out of use until then. The tool might be taken away, put in the site office and clearly labelled 'Do not use'.

Power tool checks include:

- *Look for the powered appliance testing (PAT) label*: PAT is a regular test carried out by a competent person (eg a qualified electrician) to ensure the tool is in a safe electrical condition. A sticker is placed on the tool after it has been tested. Tools that do not pass the PAT are taken out of use.

PAT testing labels

- *Cable*: Is it damaged? Is there a repair? Insulation tape may be hiding a damaged cable. Damaged cables must be replaced.

- *Casing*: Is the casing cracked? Plastic casings ensure the tool is double-insulated. This means the live parts inside are safely shielded from the user. A cracked casing will reduce the protection to the user and will require repair.

- *Guards and tooling*: Are guards in place? Is the tooling sharp?

- *Electricity supply leads*: Are they damaged? Are they creating a trip hazard? You need to place them in such a way that they do not make a trip hazard. Are they protected from damage? If they are lying on the floor with heavy traffic crossing them, they must be covered.

- *Use appropriate equipment for the size of the job*: For example, too many splitters can result in a web of cables.

- *Storage*: After use, power tools and equipment should be stored correctly. Tools must be returned to the boxes, including all the guards and parts. Cables need to be wound onto reels or neatly coiled as they can become tangled very easily.

Cable protection

Cable reel

INDUSTRY TIP

Remember, always fully unroll an extension lead before use because it could overheat and cause a fire.

FIRE

Fire needs three things to start; if just one of them is missing there will be no fire. If all are present then a fire is unavoidable:

1 *Oxygen*: A naturally occurring gas in the air that combines with flammable substances under certain circumstances.

2 *Heat*: A source of fire, such as a hot spark from a grinder or naked flame.

3 *Fuel*: Things that will burn such as acetone, timber, cardboard or paper.

The fire triangle

If you have heat, fuel and oxygen you will have a fire. Remove any of these and the fire will go out.

PREVENTING THE SPREAD OF FIRE

Being tidy will help prevent fires starting and spreading. For instance:

- Wood offcuts should not be left in big piles or standing up against a wall. Instead, useable offcuts should be stored in racks.

- Put waste into the allocated disposal bins or skips.

- Always replace the cap on unused fuel containers when you put them away. Otherwise they are a potential source of danger.

- Flammable liquids (not limited to fuel-flammable liquids) such as oil-based paint, thinners and oil must be stored in a locked metal cupboard or shed.

- Smoking around flammable substances should be avoided.

- Dust can be explosive, so when doing work that produces wood dust it is important to use some form of extraction and have good ventilation.

FIRE EXTINGUISHERS AND THEIR USES

You need to know where the fire extinguishers and blankets are located and which fire extinguishers can be used on different fires. The table below shows the different classes of fire and which extinguisher to use in each case.

Class of fire	Materials	Type of extinguisher
A	Wood, paper, hair, textiles	Water, foam, dry powder, wet chemical
B	Flammable liquids	Foam, dry powder, CO_2
C	Flammable gases	Dry powder, CO_2
D	Flammable metals	Specially formulated dry powder
E	Electrical fires	CO_2, dry powder
F	Cooking oils	Wet chemical, fire blanket

Fire blanket

CO_2 extinguisher

Dry powder extinguisher

Water extinguisher

Foam extinguisher

It is important to use the correct extinguisher for the type of fire as using the wrong one could make the danger much worse, eg using water on an electrical fire could lead to the user being electrocuted!

EMERGENCY PROCEDURES

In an emergency, people tend to panic. If an emergency were to occur, such as fire, discovering a bomb or some other security problem, would you know what to do? It is vital to be prepared in case of an emergency.

It is your responsibility to know the emergency procedures on your work site:

- If you discover a fire or other emergency you will need to raise the alarm:
  - You will need to tell a nominated person. Who is this?
  - If you are first on the scene you will have to ring the emergency services on 999.

- Be aware of the alarm signal. Is it a bell, a voice or a siren?

- Where is the assembly point? You will have to proceed to this point in an orderly way. Leave all your belongings behind, they may slow you or others down.

- At the assembly point, there will be someone who will ensure everyone is out safely and will do so by taking a count. Do you know who this person is? If during a fire you are not accounted for, a firefighter may risk their life to go into the building to look for you.

- How do you know it's safe to re-enter the building? You will be told by the appointed person. It's very important that you do not re-enter the building until you are told to do so.

Emergency procedure sign

ACTIVITY

What is the fire evacuation procedure at your workplace or place of training?

SIGNS AND SAFETY NOTICES

The law sets out the types of safety signs needed on a construction site. Some signs that warn us about danger and others tell us what to do to stay safe.

The following table describes five basic types of sign.

Type of sign	Description
Prohibition	These signs are red and white. They are round. They signify something that must *not* be done.
Mandatory 	These signs are blue. They are round. They signify something that *must* be done.

Type of sign	Description
Caution 	These signs are yellow and black. They are triangular. These give warning of hazards.
Safe condition 	These signs are green. They are usually square or rectangular. They tell you the safe way to go, or what to do in an emergency.
Supplementary 	These white signs are square or rectangular and give additional important information. They usually accompany the signs above.

Case Study: Graham and Anton

An old barn had planning passed in order for it to be converted into a dwelling.

Keith, the contractor, was appointed and the small building company turned up first thing Monday morning.

Graham, the foreman, took a short ladder off the van to access the building's asbestos and slate roof to inspect its condition. The ladder just reached fascia level. As Graham stepped off onto the roof the ladder fell away, leaving him stranded. Luckily for him, Anton the apprentice, who was sitting in the van at the time noticed what had happened and rushed over to put the ladder back up.

While inspecting the whole roof Graham found that the asbestos roof covering was rather old and had become brittle over time, especially the clear plastic roof light sections. It was also clear upon close inspection that the ridge had holes in it and was leaking water. On the slated area of the roof it was noted that many slates were loose and some of them had fallen away leaving the battens and rafters exposed which was leading to severe decay of the timbers.

It was decided that the whole roof needed to be replaced.

- Was the survey carried out safely?

- What accidents could have happened during the survey?

- What could have been done to make the whole operation safer?

- What is the builder's general view of safety?

- How would you carry out the roof work in a safe fashion?

Work through the following questions to check your learning.

1 Which of the following must be filled out prior to carrying out a site task?

   a   Invoice

   b   Bill of quantities

   c   Risk assessment

   d   Schedule

2 Which of the following signs shows you something you *must* do?

   a   Green circle

   b   Yellow triangle

   c   White square

   d   Blue circle

3 Two parts of the fire triangle are heat and fuel. What is the third?

   a   Nitrogen

   b   Oxygen

   c   Carbon dioxide

   d   Hydrogen sulphite

4 Which of the following types of fire extinguisher would best put out an electrical fire?

   a CO_2

   b   Powder

   c   Water

   d   Foam

5 Which piece of health and safety legislation is designed to protect an operative from ill health and injury when using solvents and adhesives?

   a   Manual Handling Operations Regulations 1992

   b   Control of Substances Hazardous to Health (COSHH) Regulations 2002

   c   Health and Safety (First Aid) Regulations 1981

   d   Lifting Operations and Lifting Equipment Regulations (LOLER) 1998

6 What is the correct angle at which to lean a ladder against a wall?

   a   70°

   b   80°

   c   75°

   d   85°

7 Which are the most important pieces of PPE to use when using a disc cutter?

   a   Overalls, gloves and boots

   b   Boots, head protection and overalls

   c   Glasses, hearing protection and dust mask

   d   Gloves, head protection and boots

8 Which of these is not a lifting aid?

   a   Wheelbarrow

   b   Kerb lifter

   c   Gin lift

   d   Respirator

9 Which of these is a 3 phase voltage?

   a   410V

   b   230V

   c   240V

   d   110V

10 Above what noise level must you wear ear protection?

   a   75dB (a)

   b   80dB (a)

   c   85dB (a)

   d   90dB (a)

Chapter 2
Unit 101: Principles of building construction, information and communication

Working in the building industry involves more than just the physical construction of buildings such as laying blocks, screwing timber together or soldering pipes. Building is an expensive business and for the work to progress smoothly (and on budget) the work needs to be well organised.

This involves interpreting information such as drawings, specifications and schedules. It also involves calculating quantities and dimensions and knowing how to communicate well with others.

By reading this chapter you will know about:

1 Identifying information used in the workplace.
2 Environmental considerations in relation to construction.
3 Construction of foundations.
4 Construction of internal and external walls.
5 Construction of floors.
6 Construction of roofs.
7 Communicating in the workplace.

TECHNICAL INFORMATION

This section will discuss the three main sources of technical information that are used when constructing buildings:

- working drawings and **specifications**

- schedules

- **bill of quantities**.

These are all essential information and form the contract documents (those that govern the construction of a building). All documentation needs to be correctly interpreted and correctly used. The contract documents need to be looked after and stored (filed) correctly and safely. If documents are left lying around they will become difficult to read and pages may be lost, leading to errors. The contract documents will need to be **archived** at the end of the contract, so they can be referred back to in case of any query or dispute over the work carried out or the materials used.

DRAWING SCALES

It is impossible to fit a full-sized drawing of a building onto a sheet of paper, so it is necessary to **scale** (shrink) the size of the building to enable it to fit. The building has to be shrunk in proportion; this makes it possible to convert measurements on the drawing into real measurements that can be used. Scale rules are made specifically for this purpose.

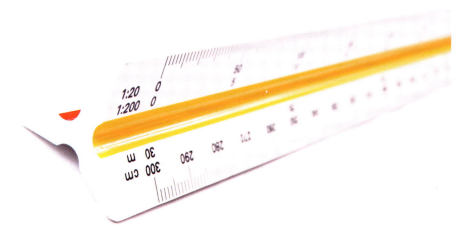

Triangular scale rule

Specification

A contract document that gives information about the quality of materials and standards of workmanship required

Bill of quantities

A document containing quantities, descriptions and cost of works and resources

Archived

Kept in storage

Scale

The ratio of the size on a drawing to the size of the real thing that it represents

INDUSTRY TIP

How do scale rules work? Let's say we are using a scale of 1:5. That means that what we draw – using the sizes on the scale rule – will be five times smaller on the drawing than the object's actual size. So, a line 30mm long will represent an object 150mm long (30 x 5 = 150).

INDUSTRY TIP

Do not scale from photocopies because these can easily become distorted in the process of photocopying.

INDUSTRY TIP

If a drawing has **dimensions**, use these instead of using a scale rule to take a measurement.

Dimension

A measurement

The British Standards Institute's BS 1192 (Drawing office practice) gives a range of standard scales that are used for various drawing types and scale rules are manufactured to meet this purpose.

British Standards Institute

The British Standards Institute (BSI) is the UK authority which develops and publishes standards in the UK

SCALES IN COMMON USE

Scale	Use
1:1	Full size (used for rods)
1:2 1:5 1:10	Building details
1:20 1:50 1:100 1:200	Plans, elevations and sections
1:200 1:500 1:1250	Site plans
1:1250 1:2500	Location plans

ACTIVITY

Work out the following:

Scale size	Scale	Actual size
10mm	1:10	100mm
25mm	1:20	a)
b)	1:50	300mm
50mm	1:200	c)

Answers: a) 500mm, b) 6mm, c) 10m

The documents these scales are used for are described on pages 49–51.

DATUM POINTS

Heights of buildings and the relative heights of components within the building are calculated from a common **datum point**. Datum points are determined by transferring a known fixed height from a bench mark. There are two types of datum point:

- A permanent Ordnance bench mark (OBM) is a given height on an Ordnance Survey map. This fixed height is described as a value, eg so many metres above sea level (as calculated from the average sea height at Newlyn, Cornwall).

- A temporary bench mark (TBM) is set up on site.

Datum point

A fixed point or height from which reference levels can be taken. The datum point is used to transfer levels across a building site. It represents the finished floor level (FFL) on a dwelling

Ordnance and temporary bench marks

ACTIVITY

Find your local OBM or your site TBM.

BASIC DRAWING SYMBOLS (HATCHINGS)

Standard symbols, also known as hatching symbols, are used on drawings as a means of passing on information simply. If all the parts of a building were labelled in writing, the drawing would soon become very crowded. Additionally, it is important to use standard symbols so that everyone can read them and it means the same to everyone. The following images are just some of the standard symbols used.

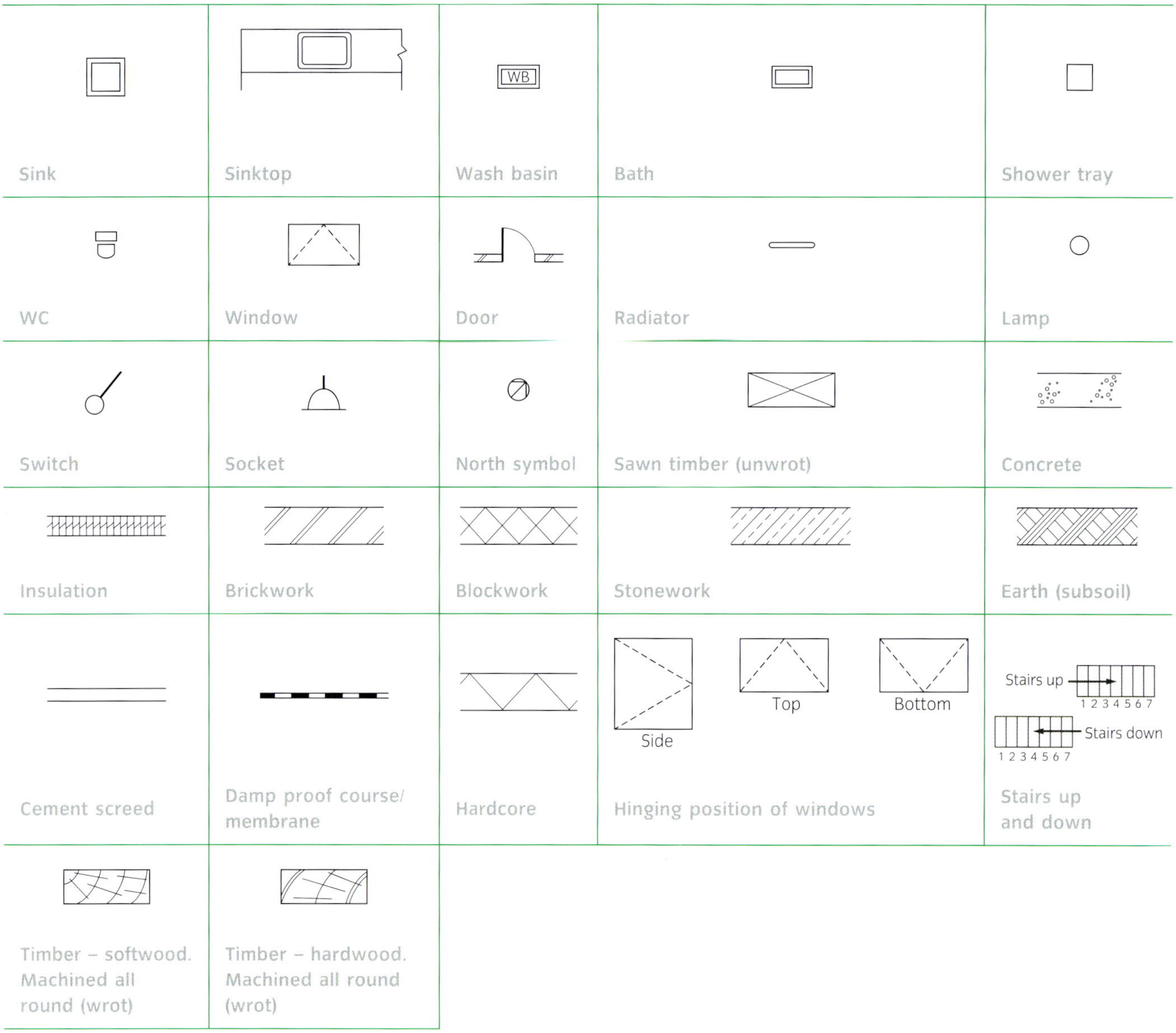

Sink
Sinktop
WB
Wash basin
Bath
Shower tray
WC
Window
Door
Radiator
Lamp
Switch
Socket
North symbol
Sawn timber (unwrot)
Concrete
Insulation
Brickwork
Blockwork
Stonework
Earth (subsoil)
Cement screed
Damp proof course/ membrane
Hardcore
Side
Top
Bottom
Hinging position of windows
Stairs up
1 2 3 4 5 6 7
Stairs down
1 2 3 4 5 6 7
Stairs up and down
Timber – softwood. Machined all round (wrot)
Timber – hardwood. Machined all round (wrot)

INFORMATION SOURCES

Type of drawing	Description
Location drawings	Usually prepared by an **architect** or **architectural technician**. Show the location of the building plot, position of the building and areas within the building. Location drawings covers all of the drawings in this table.
Location plans 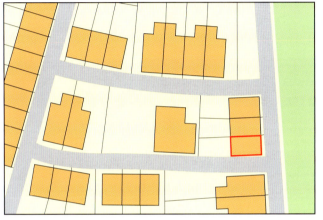 Scale 1:1250	Show the proposed development in relation to its surrounding properties. The scales used are 1:1250 or 1:2500. Very little detail is available from this type of plan. The direction North is usually shown.
Block plans (Site plans) 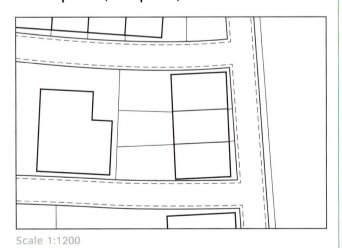 Scale 1:1200	Show the plot in more detail, with drain runs, road layouts and the size and position of the existing building (and any extensions proposed) in relation to the property boundary. A scale of 1:500 or 1:200 is used.

Architect

A trained professional who designs a structure and represents the client who wants the structure built. They are responsible for the production of the working drawings. They supervise the construction of buildings or other large structures

Architectural technician

A draftsperson who works in an architectural practice

Type of drawing	Description
Floor plans 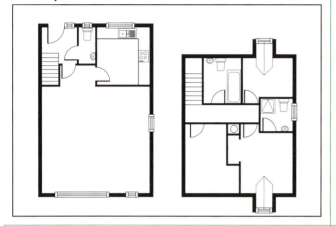	Show the positioning of walls, size of rooms along with the positioning of elements within the building such as units.
Elevations 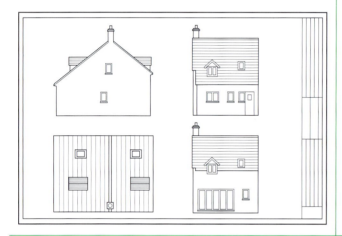	Show a building from a particular side and show the positioning of features such as doors and windows.
Sections 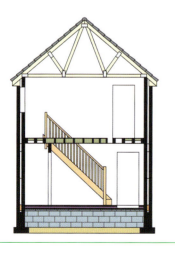	Show in greater detail what the section of a component looks like and how it might fit in relation to another component. A typical example would be a cross-section of a window showing the size of the features and how it fits together. Using these drawings it is possible to determine the positions of rooms, windows, doors, kitchen units and so on. Elevations are shown. These drawings are more detailed, and are often scaled to provide construction measurements. Scales used are 1:200, 1:100 and 1:50, 1:10, 1:5 and 1:1. A scale of 1:1 is full size.

Type of drawing	Description
Construction drawings (Detail drawings)	Show details of construction, normally as a cross-section.

Detail showing typical exterior corner detail

External walls
Exterior cladding
Breather membrane paper
Exterior cladding
Wall plate stud
Breather membrane paper
Vapour control membrane on the inside of the timber frame

SPECIFICATIONS

A specification accompanies the working drawings. They give further information that cannot be shown on the drawings. The drawings need to be clear and not covered in notes. A specification would include information such as:

- the colour of paint required

- a specific timber species

- the brick type required

- the plaster finish required.

They are prepared by construction professionals such as architects and building services engineers. They can be produced from previous project specifications, in-house documents or master specifications such as the National Building Specification (NBS). The NBS is owned by the Royal Institute of British Architects (RIBA).

Example of a specification

COMPONENT RANGE DRAWINGS

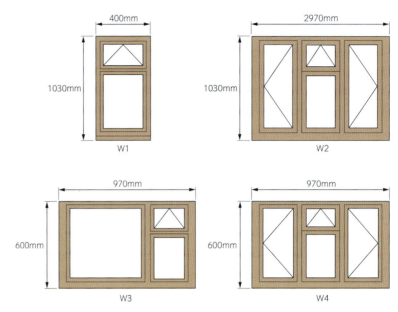

Component range drawing of windows

A component range drawing shows the range of components available from a manufacturer. It includes:

- sizes available

- coding for ordering purposes

- availability (whether it can be bought off-the-shelf or if pre-ordering is required).

Availability is particularly important when planning delivery dates. Schedules reference this type of drawing.

SCHEDULES

A schedule is used to record repeated design information that applies to a range of components or fittings, such as:

- windows

- doors

- kitchen units

- joinery fittings.

A schedule is mainly used on bigger sites where there are multiples of several designs of houses, with each type having different components and fittings. It avoids the wrong component or fitting being put in the wrong house.

A schedule is usually used in conjunction with a component range drawing and a floor plan.

In a typical plan, the doors and windows are labelled D1, D2, W1, W2 etc. These components would be included in the schedule, which would provide additional information on them, for example see the following schedule.

Master Internal Door Schedule							
Ref:	Door size	S.O. width	S.O. height	Lintel type	FD30	Self closing	Floor level
D1	838x1981	900	2040	BOX	Yes	Yes	GROUND FLOOR
D2	838x1981	900	2040	BOX	Yes	Yes	GROUND FLOOR
D3	762x1981	824	2040	BOX	No	No	GROUND FLOOR
D4	838x1981	900	2040	N/A	Yes	No	GROUND FLOOR
D5	838x1981	900	2040	BOX	Yes	Yes	GROUND FLOOR
D6	762x1981	824	2040	BOX	Yes	Yes	FIRST FLOOR
D7	762x1981	824	2040	BOX	Yes	Yes	FIRST FLOOR
D8	762x1981	824	2040	N/A	Yes	No	FIRST FLOOR
D9	762x1981	824	2040	BOX	Yes	Yes	FIRST FLOOR
D10	762x1981	824	2040	N/A	No	No	FIRST FLOOR
D11	686x1981	748	2040	N/A	Yes	No	SECOND FLOOR
D12	762x1981	824	2040	BOX	Yes	Yes	SECOND FLOOR
D13	762x1981	824	2040	100 HD BOX	Yes	Yes	SECOND FLOOR
D14	686x1981	748	2040	N/A	No	No	SECOND FLOOR

Example of a schedule

BILL OF QUANTITIES

A bill of quantities is produced by the quantity surveyor and describes everything that is required for the job based on the drawings, specification and schedules. A bill of quantities contains the following information:

- *Preliminaries*: General information including the client and architect, details of the work and descriptions of the site.

- *Preambles*: Like the specification, this outlines the quality and description of materials and workmanship, etc.

- *Measured quantities*: A description of how each task and material is to be measured, with measurements in metres (linear and square), hours, litres, kilogrammes and the number of components required.

The completed document is sent out to contractors who will then price the work and enter the costs into the blank spaces. The bill of quantities ensures that all the contractors are pricing for the job using the same information.

BILL OF QUANTITIES

(Assuming Civil Engineering Standard Method of Measurement (CESSM3) is used.)

Number	Item description	Unit	Quantity	Rate	Amount £	p
	CLASS A: GENERAL ITEMS					
	Specified Requirements					
	Testing of Materials					
A250	Testing of recycled and secondary aggregates	sum				
	Information to be provided by the Contractor					
A290	Production of Materials Management Plan	sum				
	Method Related Charges					
	Recycling Plant / Equipment					
A339.01	Mobilise; Fixed	sum				
A339.02	Operate; Time-Related	sum				
A339.03	De-mobilise; Fixed	sum				
	CLASS D: DEMOLITION AND SITE CLEARANCE					
	Other Structures					
D522.01	Other structures; Concrete;	sum				
D522.02	Grading / processing of demolition material to produce recycled and secondary aggregates	m³	70			
D522.03	Disposal of demolition material offsite	m³	30			
	CLASS E: EARTHWORKS					
	Excavation Ancillaries					

Bill of quantities

WORK SCHEDULES

It is very important indeed that the progress of work is planned out. A work schedule or programme of work is an easy way of showing what work is to be carried out and when. This is usually shown in the form of a bar chart called a gantt chart. The chart lists the tasks that need to be done on the left-hand side and shows a timeline across the top. The site manager or trade supervisors can quickly tell from looking at this chart:

- if work is keeping to schedule

- what materials, equipment, labour are required

- when they are required.

Materials very often have a lead-in time and so cannot be delivered immediately. These need to be ordered and delivered at the correct time. Labour planning is also required as the trades may be working elsewhere when needed.

INDUSTRY TIP

Use of a planning document such as a gantt chart will reduce waste and ensure effective use of labour.

Task	Time (days)						
	1	2	3	4	5	6	7
Prepare the ground	�	▰					
Spread foundations			▰	▰			
Lay cables for services				▰	▰		
Build walls up to DPC						▰	▰
Proposed time in green							

Gantt chart

CALCULATING QUANTITIES FOR MATERIALS

Calculations are required throughout the building process. It is important that these calculations are accurate, as mistakes can be very expensive. Several factors can have an impact on cost. Underestimating:

- the amount of materials required

- how much they cost

- how long it will take to complete a job

can lose the company a lot of money. It could also lead to the company gaining a bad reputation for not being able to complete a job on time and in budget.

Materials are usually better priced if bought in bulk, whereas a buy-as-you go approach can cost more. Consider these points when buying materials:

- Is there sufficient storage room for delivered materials?

- Is there a risk of the materials being damaged if there is nowhere suitable to store them or if they are delivered too early?

- Will it be a problem to obtain the same style, colour or quality of product if they are not all ordered at the same time?

- Will over-ordering cause lots of wastage ?

These and many other considerations will help determine when and in what quantity materials are ordered.

Some wastage is unavoidable. Allowances must be made for wastage, eg cut bricks that cannot be re-used, short ends of timber, partly full paint cans. Up to 5% waste is allowed for bricks and blocks and 10% for timber and paint.

It may be that all the materials are ordered by the office or supervisory staff, but you still need to know how to recognise and calculate material requirements. Deliveries have to be checked before the delivery note is signed and the driver leaves. Any discrepancies in the type or quantity of materials, or any materials that have arrived damaged, must be recorded on the delivery note and reported to the supervisor. Any discrepancies will need to be followed up and new delivery times arranged.

You must be able to identify basic materials and carry out basic calculations. You will often have to collect sufficient materials to carry out a particular operation. Being able to measure accurately will mean you can make the most economic use of materials and therefore reduce waste.

Deliveries must be checked before signing the delivery note

UNITS OF MEASUREMENT

The construction industry uses metric units as standard; however you may come across some older measures called imperial units.

Units for measuring	Metric units	Imperial units
Length	millimetre (mm) metre (m) kilometre (km)	inch (in) or " eg 6" (6 inches) foot (ft) or ' eg 8' (8 foot)
Liquid	millilitre (ml) litre (l)	pint (pt)
Weight	gramme (g) kilogramme (kg) tonne (t)	pound (lb)

ACTIVITY

Look online to find out:
- What other imperial units are still commonly used?
- How many millimetres are there in an inch?
- How many litres are there in a gallon?

Units for measuring	Quantities	Example
Length	There are 1000mm in 1m. There are 1000m in 1km.	1mm x 1000 = 1m 1m x 1000 = 1km 6250mm can be shown as 6.250m 6250m can be shown as 6.250km
Liquid	There are 1000ml in 1l.	1ml x 1000 = 1l
Weight	There are 1000g in 1kg. There are 1000kg in 1t.	1g x 1000 = 1kg 1kg x 1000 = 1t

CALCULATIONS

Four basic mathematical operations are used in construction calculations.

ADDITION

The addition of two or more numbers is shown with a plus sign (**+**).

Example 1

A stack of bricks is 3 bricks long and 2 bricks high. It contains 6 bricks.

$$3 + 3 = \mathbf{6}$$

More examples:

$$5 + 2 = \mathbf{7}$$

$$19 + 12 = \mathbf{31}$$

$$234 + 105 = \mathbf{339}$$

Pallet of bricks

SUBTRACTION

The reduction of one number by another number is shown with a minus sign (**–**).

Example 1

A pallet containing 100 bricks is delivered on site, but you only need 88 bricks. How many are left over?

$$100 - 88 = \mathbf{12}$$

More examples:

$$5 - 2 = \mathbf{3}$$

$$19 - 12 = \mathbf{7}$$

$$234 - 105 = \mathbf{129}$$

MULTIPLICATION

The scaling of one number by another number is shown with a multiplication sign (**x**).

Example 1

A stack of bricks is 3 bricks long and 2 bricks high. It contains 6 bricks.

$$3 \times 2 = \mathbf{6}$$

More examples:

$$19 \times 12 = \mathbf{228}$$

$$234 \times 10 = \mathbf{2340}$$

$$234 \times 105 = \mathbf{24,570}$$

In the last example, the comma (,) is used to show we are in the thousands. In words we would say, twenty four thousand, five hundred and seventy.

DIVISION

Sharing one number by another number in equal parts (how many times it goes into the number) is shown with a division sign (÷).

Example 1

$$5 \div 2 = \mathbf{2.5}$$

$$36 \div 12 = \mathbf{3}$$

$$600 \div 4 = \mathbf{150}$$

LINEAR LENGTH

Linear means how long a number of items would measure from end to end if laid in a straight line. Examples of things that are calculated in linear measurements are:

- skirting board
- lengths of timber
- rope
- building line
- wallpaper.

We use this form of measurement when working out how much of one of the materials listed above we need, eg to find out how much

Skirting boards are calculated using linear measurements

A joiner measuring a room

skirting board is required for a room. First, we need to measure the **perimeter** (sides) of a room. To find the linear length we add the length of all four sides together. This can be done in two ways: adding or multiplying.

Example 1

A site carpenter has been asked how many metres of skirting are required for the rooms below.

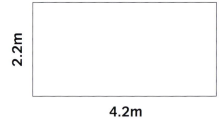

2.2m

4.2m

They can add all the sides together:
2.2 + 4.2 + 2.2 + 4.2 = 12.8m

Or, they can multiply each side by 2, and add them together:
(2.2 x 2) + (4.2 x 2) = 12.8m

Either way, **12.8m** is the correct answer.

Example 2

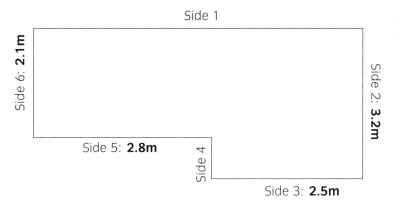

Side 1

Side 6: **2.1m**

Side 2: **3.2m**

Side 5: **2.8m**

Side 4

Side 3: **2.5m**

To work out the perimeter of this room we need to add all the sides together. In this example each side has been given a reference number, so all we need to do is add all the sides together, like this:

side 1 (side 3 + side 5) + side 2 + side 3 + side 4 (side 2 - side 6) + side 5 + side 6

Now, let's show the working out: 2.8 + 2.5 + 3.2 + 2.5 + 3.2 - 2.1 + 2.8 + 2.1 = 17m

The amount of skirting board required is **17m**.

Now let's put some door openings in. This symbol represents an opening.

Example 3

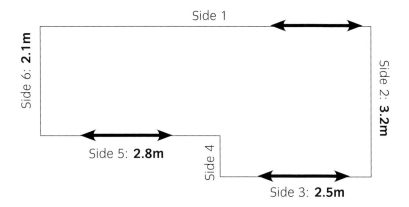

On side 1 there is an opening 0.9m wide, on side 3 there is an opening 1.5m wide and on side 5 there is an opening 2.1m wide.

We know from Example 2 that the perimeter of the room is 17m. We now need to remove the openings. Skirting board will not be needed for the openings.

Step 1

Add together the lengths of the three combined openings:

0.9 + 1.5 + 2.1 = 4.5m

Step 2

Deduct this from 17m:

17 − 4.5 = 12.5m

The linear length of skirting board required is 12.5m.

Step 3

However, this calculation does not take into account any waste. We would normally add 10% extra to allow for waste:

12.5 + 10% = 12.5 + 1.25 = 13.75m

The total amount of skirting board required is **13.75m**.

PERCENTAGES

An easy way to find a percentage (%) of a number is to divide the number by 100 and then multiply it by the percentage you require.

For example:

Increase 19m by 12%

19 ÷ 100 = 0.19

0.19 x 12 = 2.28

19 + 2.28 = 21.28m

Total required **21.28m**.

AREA

To find out how much material is required to cover a surface such as a **floor** or wall you need to calculate its area. Area is the measurements of a two dimensional surface, eg the surface of floors, walls, glass or a roof.

To find the area of a surface you need to multiply its length by its width (L x W) or one side by the other. This will give you an answer which is expressed in square units (^{2}). For example, mm^2, m^2 or km^2.

Floors

The structured layers of a building, eg ground floor, first floor, second floor

Example 1

A bricklayer has been asked to work out the area of the floors below.

Side 1: **2.2m**

Side 2: **4.4m**

side 1 x side 2 = floor area

2.2 x 4.4 = 9.68m^2

The total floor area is **9.68m²**.

Irregularly shaped areas can be calculated by breaking up the area into sections that can be worked out easily, and then adding them together.

Example 2

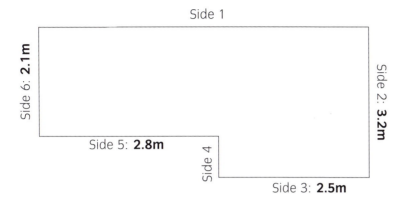

Irregularly shaped rooms can be split into sections to calculate the area

Step 1

Divide the area into two parts, and then calculate the area of each part. The easiest way to do this is to divide it into two smaller sections:

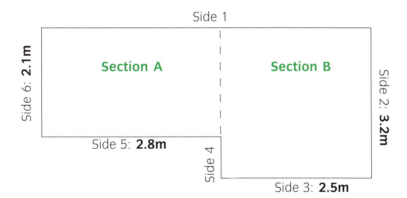

Step 2

Work out the area of section A and section B:

section A: 2.1 x 2.8 = 5.88m²

section B: 2.5 x 3.2 = 8m²

Step 3

Add the areas of section A and section B together:

section A + section B = total floor area

5.88 + 8 = 13.88m²

The total floor area is **13.88m²**.

A tiler tiling a floor

ACTIVITY

Find the area of the following measurements:

1 2.1m x 2.4m
2 0.9m x 2.7m
3 250mm x 3.4m

Answers: 1) 5.04m, 2) 2.43m, 3) 0.85m

Now let's say the floor requires tiling. The tiler needs to calculate the number of floor tiles required.

Example 3

The size of each floor tile is 305mm x 305mm. We can also show this as 0.305m x 0.305m.

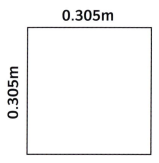

How many floor tiles are required for the floor area in Example 2? The total floor area is 13.88m².

Step 1

Calculate the area of one tile. As the floor area is given in m², we need to calculate the size of the tile in the same unit, ie m².

0.305 x 0.305 = 0.093m²

Step 2

Now you need to divide the total floor area by the area of one tile to find out the total number tiles required.

total floor area ÷ area of one tile = total number of tiles

13.88 ÷ 0.093 = 149.247 tiles

This number is rounded up to the next full tile, so a total of 150 floor tiles are required.

Step 3

However, this total does not allow for any waste.

Add 5% to allow for waste:

150 + 5% = 158 tiles (to the next full tile)

Let's look at the working out:

150 ÷ 100 = 1.5 tiles (this is 1%)

1.5 x 5 = 7.5 tiles (this is 5%)

5% of 150 tiles, rounded up to the next full tile, is 8 tiles.

Therefore **158 tiles** are required.

AREA OF A TRIANGLE

Sometimes you will be required to work out an area that includes a triangle.

Example 1

A painter has been asked to work out how much paint will be needed to paint the front of this house.

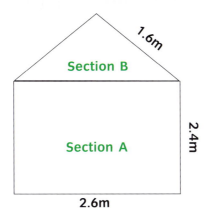

A decorator measuring a room

Step 1

Divide the area up into a rectangular section (section A) and a triangular section (section B).

Step 2

Find the area of section A:

2.4 x 2.6 = 6.24m²

The area of section A is 6.24m².

Step 3

Find the area of section B

The area of a triangle can be found by multiplying the base by the height, then dividing by 2.

(base x height) ÷ 2 = area

2.6 x 1.6 = 4.16

4.16 ÷ 2 = 2.08m²

The area of section B is 2.08m².

Step 4

area of section A + area of section B = total wall area

6.24 + 2.08 = 8.32m²

The total wall area is **8.32m²**.

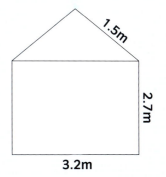

Now let's look at the simple triangle below. It has three sides, A, B and C. Pythagorean theorem tells us that in a right-angled triangle the **hypotenuse** is equal to the sum of the square of the lengths of the two other sides, in other words $a^2 + b^2 = c^2$. In this example the hypotenuse is side C.

Using the Pythagorean theorem we can work out the length of any side.

Example 1

If side A is 3m long and side B is 4m long, what is the length of side C?

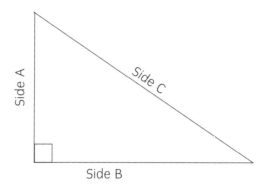

$3 \times 3 = 9$

$4 \times 4 = 16$

$9 + 16 = 25$

$\sqrt{25} = 5$

($\sqrt{}$ means square root, or a number that is multiplied by itself, in this case $5 \times 5 = 25$)

Side C is **5m** long.

Hypotenuse

The longest side of a right-angled triangle. It is always opposite the right angle

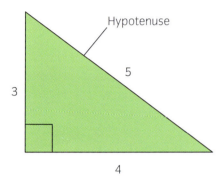

The hypotenuse

INDUSTRY TIP

If a triangle has a small square in the corner, this shows you the corner is a right angle.

Example 2

A joiner has been asked to work out the length of a roof (side C).

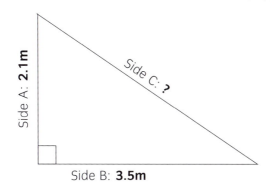

2.1 x 2.1 (side A) = 4.41

3.5 x 3.5 (side B) = 12.25

4.41 + 12.25 = 16.66

$\sqrt{16.66}$ = 4.08m

The length of side C is **4.08m**.

Example 3

A bricklayer needs to find the rise of a roof (side A).

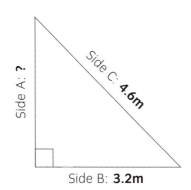

3.2 x 3.2 (side B) = 10.24

4.6 x 4.6 (side C) = 21.16

21.16 − 10.24 = 10.92

$\sqrt{10.92}$ = 3.30m

The length of side A is **3.3m**.

Use Pythagorean theorem to answer following questions:

1   What is the length of side B?

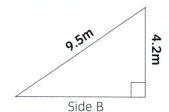

2   What is the length of side C?

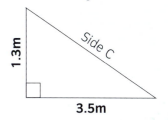

Answers: 1) 8.5m², 2) 3.73m

PERIMETERS AND AREAS OF CIRCLES

Circumference

The distance around the edge of a circle

Sometimes you are required to find the perimeter or **circumference** of a circle.

circumference of a circle = π x **diameter**

$$C = \pi d$$

π (or 'pi') is the number of times that the diameter of a circle will divide into the circumference.

π = 3.142

Diameter

The length of a straight line going through the centre of a circle connecting two points on its circumference

This is equal to the number of diameters in one revolution of a circle. It is the same for any sized circle.

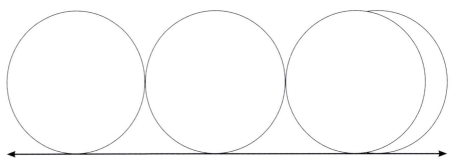

There are 3.142 diameters in one complete revolution

Example 1

A joiner is making a circular window that has a diameter of 600mm. Its circumference is:

0.600 x 3.142 = **1.885m**

The diameter of a circle from a given circumference is:

diameter = circumference ÷ π

Example 2

A window has a circumference of 2.250m. Its diameter is:

2.250 ÷ 3.142 = **0.716m** (or 716mm)

Radius

The length of a line from the centre to a point on the circumference of a circle. It is exactly half the length of the diameter

The area of a circle is found by:

area of a circle = π x **radius²** (radius is equal to half the diameter)

Example 3

A painter needs to paint a circle that is 1.2m in diameter and is required to find the area of the circle to enable them to order the correct quantity of paint.

1.2 ÷ 2 = **0.600** (the radius)

3.142 x 0.600² = **1.13m²**

VOLUME

The volume of an object is the total space it takes up, eg a tin of paint, a foundation for a wall or the capacity of a concrete mixer, and is shown as m³ (cubic metres). To find the volume of an object you must multiply length by width by height.

volume = length x width x height

Example 1

Each side of this cube is 1m. The total space it takes up is 1m³

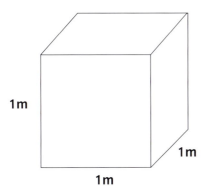

1m x 1m x 1m = **1m³**

Example 2

A bricklayer has been asked to work out how many m³ of **concrete** is required for a strip foundation. The size of the foundation is 3.2m long, 0.600m wide and 0.900m deep.

length x width x height = volume

3.2 x 0.600 x 0.900 = 1.728m³

The volume of concrete needed for the strip foundation is **1.728m³**.

A bricklayer taking levels

Concrete

Composed of cement, sand and stone, of varying size and in varying proportions

To work out the volume of a cylinder:

$$\text{volume} = \pi r^2 h \ (\pi \times r^2 \times h)$$

ACTIVITY

A bricklayer has been given two tasks:

1 Measure the volume of a strip foundation measuring 4.250m long, 1.1m wide and 1m deep.
2 Find the volume of four pile foundations (see page 80) each measuring 2.5m deep, with a diameter of 0.9m.

Work out the answers to the tasks.

Answers: 1) 4.675m³, 2) 6.36m²

Example 3

A joiner has a tin of presevative and needs to know its volume. The tin has a diameter of 250mm and a height of 700mm.

$\pi r^2 h \ (\pi \times r^2 \times h) = \text{volume}$

The radius (r) is half the diameter:

$250 \div 2 = 125\text{mm}$

$3.142 \times 0.125^2 \times 0.700 = 0.034\text{m}^3$

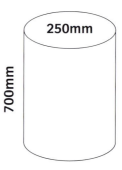

The volume of the tin of paint is **0.034m³**.

COMMUNICATION

Good communication is vital to the smooth running of any building project.

Communication involves sharing thoughts, information and ideas between people. For communication to be effective, information must be:

- given in a clear way
- received without misunderstanding.

It has been said that to be a good communicator it is just as important to be a good listener as it is to be a good speaker! Good communication leads to a safer and more efficient workplace, not to mention helping to maintain a pleasant working environment.

Most sites will have policies and procedures in place that govern the chain of command and communication between supervisory staff and workers.

ACTIVITY

A customer has asked for the best steps to take before painting the skirting board in their new home. You have been asked to reply to the customer and give advice on the best way for them to do this.

Decide on the best form of communication and list all the information you should give along with the stages they should follow.

WRITTEN COMMUNICATION

There are many methods of communication within the building industry. In this chapter we have discussed drawings, schedules and specifications etc. The architect uses these methods to communicate details about the building to the team who will **tender** for and erect the building.

Communication is usually electronic via email (with or without attachments) or through intranet sites. Drawings are very commonly distributed in electronic formats which are printed on to paper when required. Messages are often given via text.

Sometimes communication will be via a memorandum (memo), a written form of communication with a message.

Site rules, risk assessments and method statements (see Chapter 1) communicate safety information.

Tender

The process of supplying the client with a fixed quotation for the work

INDUSTRY TIP

Messages that are passed on by word of mouth are open to interpretation, so written messages often can be more clear.

SITE PAPERWORK

Communication on site is aided by the use of paperwork and without it no building site could operate. It is an important method of communication between operatives and supervisory staff, builders, architects and clients.

Type of paperwork	Description
Timesheet **Timesheet** Employer: CPF Building Co.　Employee Name: Louise Miranda　Week starting: 17/6/13 Date: 21/6/13 	Day	Job/Job Number	Start Time	Finish Time	Total Hours	Overtime
Monday	Penburthy, Falmouth 0897	9am	6pm	8	
Tuesday	Penburthy, Falmouth 0897	9am	6pm	8	
Wednesday	Penburthy, Falmouth 0897	8.30am	5.30pm	8	
Thursday	Trelawney, Truro 0901	11am	8pm	8	2
Friday	Trelawney, Truro 0901	11am	7pm	8	1
Saturday	Trelawney, Truro 0901	9am	1pm	4	
Totals				40	7 | Employee's signature:_____ Supervisor's signature: _____	Used to record the hours completed each day, and is usually the basis on which pay is calculated. Timesheets also help to work out how much the job has cost in working hours, and can give information for future estimating work when working up a tender.

Type of paperwork	Description
Job sheet 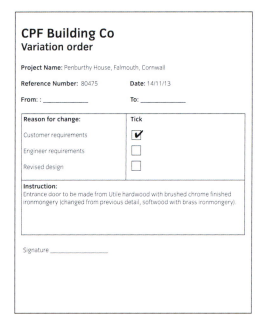	Gives details of a job to be carried out, sometimes with material requirements and hours given to complete the task.
Variation order **Confirmation notice** **Architect's instruction**	Sometimes alterations are made to the contract which changes the work to be completed, eg a client may wish to move a door position or request a different brick finish. This usually involves a variation to the cost. This work should not be carried out until a variation order and a confirmation notice have been issued.

Type of paperwork	Description
Requisition order	Filled out to order materials from a supplier or central store. These usually have to be authorised by a supervisor before they can be used.

CPF Building Co
Requisition order

Supplier Information: Construction Supplies Ltd   **Date:** 9/12/13

Contract Address/Delivery Address: Penburthy House, Falmouth, Cornwall

Tel number: 0207294333

Order Number: 26213263CPF

Item number	Description	Quantity	Unit/Unit Price	Total
X22433	75mm 4mm gauge countersunk brass screws slotted	100	30p	£3
YK7334	Brass cups to suit	100	£5	£500
V23879	Sadikkens water based clear varnish	1 litre	£20.00	£20.00
Total:				£523.00

Authorised by: Denzil Penburthy

Delivery note	Accompanies a delivery. Goods have to be checked for quantity and quality before the note is signed. Any discrepancies are recorded on the delivery note. Goods that are not suitable (because they are not as ordered or because they are of poor quality) can be refused and returned to the supplier.

Construction Supplies Ltd
Delivery note

Customer name and address:
CPF Building Co
Penburthy House
Falmouth
Cornwall

Delivery Date: 16/12/13
Delivery time: 9am

Order number: 26213263CPF

Item number	Quantity	Description	Unit Price	Total
X22433	100	75mm 4mm gauge countersunk brass screws slotted	30p	£3
YK7334	100	Brass cups to suit	£5	£500
V23879	1 litre	Sadikkens water based clear varnish	£20	£20

Subtotal	£530.00
VAT	20%
Total	£636.00

Discrepancies: ....

Customer Signature: ....

Print name: ....

Date: ....

Type of paperwork	Description
Delivery record 	Every month a supplier will issue a delivery record that lists all the materials or hire used for that month.
Site diary 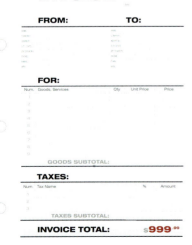	This will be filled out daily. It records anything of note that happens on site such as deliveries, absences or occurrences, eg delay due to the weather.
Invoices	Sent by the supplier. They list the services or materials supplied along with the price the contractor is requested to pay. There will be a time limit in which to pay. Sometimes there will be a discount for quick payment or penalties for late payment.

VERBAL COMMUNICATION

Often, managers, supervisors, work colleagues and trades communicate verbally. This can be face to face or over a telephone. Although this is the most common form of communication, it is also the most unreliable.

Mistakes are often made while communicating verbally. The person giving the information might make an error. The person receiving the information might misunderstand something because the information is unclear or it is noisy in the background, or because they later forget the details of the conversation.

Confusion can be minimised by recording conversations or by using a form of written communication. If there is a record it can be used for future reference and help to clear up any misunderstandings.

TAKING A TELEPHONE MESSAGE

It is a good idea to take down details of telephone calls and many companies provide documentation for this purpose. When taking a message it is important to record the following details:

- *Content*: This is the most important part of the message – the actual information being relayed. Take and write down as many details as possible.

- *Date and time*: Messages are often time sensitive, and may require an urgent response.

- *Who the message is for*: Ensure the person gets the message by giving it to them or leaving it in a place where they will find it.

- *Contact name and details*: Write down the name of the person leaving the message, and how to get back to them with a response.

UNACCEPTABLE COMMUNICATION

When communicating, it is very important to stay calm. Think about what you are going to say. An angry word will often encourage an angry response. However, keeping calm and composed will often diffuse a stressful situation. A shouting match rarely ends with a good or productive result.

There are several types of communication that are unacceptable and could result in unemployment. Unacceptable communication includes:

- aggressive communication such as swearing or using inappropriate hand gestures

An operative taking notes during a phone call

- racist or sexist comments or gestures

- showing prejudice against people with disabilities.

This type of behaviour shows a lack of respect for others and does not create a safe or pleasant working environment. It will also give your company a poor image if customers see or hear this behaviour. Acting in this way is likely to result in trouble for you and your employer and could even result in a **tribunal** and loss of employment.

Tribunal

A judgement made in court

KNOWLEDGE OF THE CONSTRUCTION INDUSTRY AND BUILT ENVIRONMENT

Buildings come in a wide variety of types in relation to appearance and methods of construction. Despite the variety of buildings, they all have design features in common. In this section we will discuss various parts of buildings and their purpose.

We will also discuss sustainable construction – how buildings can be designed to sit better within the environment, with lower pollution levels and energy requirements both during the building process and when in use.

A house with solar panels

FOUNDATIONS

Foundations serve as a good base on which to put the building. They need to be capable of carrying the weight of the building and any further load that may be put upon it. These are known as **dead loads** and **imposed loads**.

Foundations must be designed to resist any potential movement in the ground on which the building will sit. Ground conditions can vary widely. Soil samples are taken to help decide on the type of foundation to use. This usually takes the form of bore holes dug or drilled around the site. These samples are sent away for testing in a laboratory. The results will identify:

- the soil condition (clay or sandy)

- the depth of the soil

- the depth of the water table

- if any contaminations are present.

The soil condition is important: clay soil drains poorly and can move if it gets waterlogged or dries out completely. Sandy soils drain very well, but can become unstable. A foundation that is suitable for the ground type and load of the building will be designed.

Foundation

Used to spread the load of a building to the sub soil

Dead load

The weight of all the materials used to construct the building

Imposed load

Additional loads that may be placed on the structure, eg people, furniture, wind and snow

INDUSTRY TIP

The type of foundation to be used will usually be decided by the architect and a structural engineer and will be the result of tests.

TYPES OF FOUNDATION

Different types of structures, such as detached houses, high rise and low rise buildings, will all require different types of foundation.

High rise building

Low rise building

Detached house

OUR HOUSE

What type of foundation does the building you are sitting in have? How can you tell? Why was that foundation type chosen? Look at the foundations used in 'Our House' as a further guide.

STRIP FOUNDATIONS

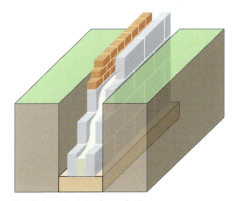

Traditional strip foundation

A strip foundation is the traditional type of foundation used for residential developments (ordinary houses). It is formed by digging a trench to the required width and depth as determined by the soil conditions and the weight of the structure. It is either filled with concrete or a layer of concrete is poured into the bottom. This layer must be a minimum of 150mm thick and is commonly 225mm thick.

Footings are brought up to the level of the **damp proof course** (DPC) using concrete blocks or bricks. These are set out from the centre of the strip of concrete in order to spread the weight evenly. A variety of specialist bricks and blocks are used for this purpose. They need to be able to resist water penetration and therefore frost damage.

Footings

The substructure below ground level. These are projecting courses at the base of a wall

Damp proof course (DPC)

A layer of plastic that prevents damp rising up through a wall needs to be positioned at least 150mm above ground level

Engineering brick

Trench block

It can be economical to fill the trench up to the top with concrete rather than build a substructure – this is known as trench fill. Sometimes it is necessary to build on the edge of the concrete – this is known as an eccentric foundation.

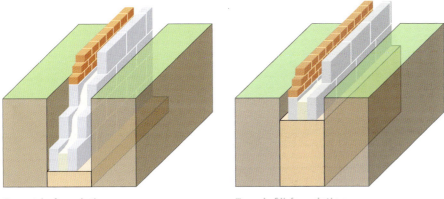

Eccentric foundation   Trench fill foundation

WIDE STRIP FOUNDATIONS

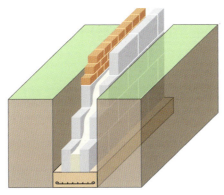

Wide strip foundation

A wide strip foundation is very similar to strip foundation in most of its aspects. The main difference between the two is that a wide strip foundation has steel reinforcements placed within the concrete. The steel gives considerably more strength to the foundation and enables greater loads to be placed on it. Without the steel reinforcements the foundation would need to be much deeper and would need vast amounts of concrete.

PAD FOUNDATIONS

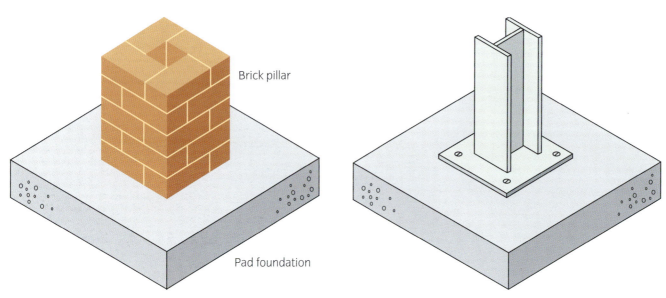

Brick pillar

Pad foundation

Pad foundation

Pad foundation with bolts

A pad foundation is used to support a point load such as a column in a steel-framed building. This type of foundation often has bolts set into the top ready for fixing the steel.

PILE FOUNDATIONS

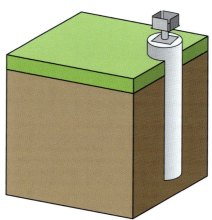

A cylindrical foundation

Deep piles are used to transfer the load through unsuitable soil layers into the harder layers of ground below, even down to rock if required (known as end bearing). Some piles use **friction** to provide support. This is known as skin friction. Tall buildings (and especially narrow buildings such as chimneys or towers) have large lateral forces due to side winds and pile foundations resist these forces well.

INDUSTRY TIP

Foundations are made from concrete. Concrete is made from fine and coarse aggregate (crushed stone) and cement mixed with water. Water reacts with the cement causing it to harden and lock the aggregates together. Concrete is very strong under compression (when weight is put upon it) but is weak when it is pulled (put under tension); therefore steel rods are cast into it to make it stronger.

Friction

Resistance between the surface of the concrete foundation and the soil around it

---

RAFT FOUNDATIONS

A raft foundation is often laid over an area of softer soil that would be unsuitable for a strip foundation. A raft foundation is a slab of concrete covering the entire base of the building; it spreads the weight of the building over a wider area but still maintains a deeper base around the load bearing walls.

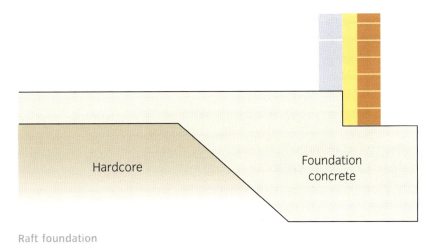

Hardcore

Foundation concrete

Raft foundation

FLOORS

Floors can be divided into two main categories:

- ground floors
- upper floors.

Floors are required to be load bearing, and there is a wide variety of construction methods depending on the type of building and potential load that will be imposed upon the floor. Floors also may need to prevent:

- heat loss
- transfer of sound
- moisture penetration.

GROUND FLOORS

These may be either solid ground floors or suspended floors.

SOLID FLOORS

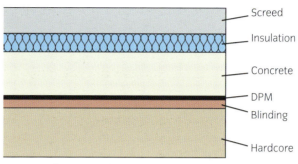

Screed
Insulation
Concrete
DPM
Blinding
Hardcore

Concrete floor

Hardcore

A mixture of crushed stone and sand laid and compacted to give a good base for the concrete

Damp proof membrane (DPM)

An impermeable layer that prevents damp coming up through the floor. A layer of sand known as blinding is placed below the DPM to prevent any sharp stones below piercing the membrane when the concrete is poured

Insulation

Materials used to retain heat and improve the thermal value of a building, may also be used for managing sound transfer

Solid concrete floors are laid upon **hardcore** and have a **damp proof membrane** (DPM) built into them to prevent damp coming up through the floor. **Insulation** is also laid into the floor to reduce heat loss. It is important that the insulation is not affected by the high water content of the wet concrete when being poured.

Steel reinforcement can also be used within the concrete to increase strength and reduce cracks.

HOLLOW AND SUSPENDED FLOORS

Upper floors, and some ground floors, are suspended or hollow meaning that instead of resting on the ground beneath, the load is transferred via beams to the walls. Two types of beam used are Posibeam and I-beam. Timber joists are usually covered with either chip board or solid timber floor boards.

Concrete with steel reinforcement

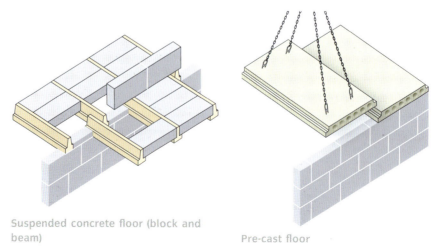

Suspended concrete floor (block and beam)

Pre-cast floor

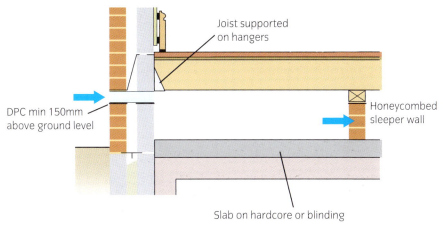

Joist supported on hangers

DPC min 150mm above ground level

Honeycombed sleeper wall

Slab on hardcore or blinding

Suspended wood floor

Posibeam

I-beam

UPPER FLOORS

In most domestic dwellings timber floor joists are used following the same principle as timber ground floors, while in large commercial and industrial buildings solid concrete floors are used.

WALLS

Walling for a building can usually be divided in two categories:

- external
- internal.

Walling can be load or non-load bearing. Load bearing walls carry the weight of the floors and roof and transfer this weight down to the foundations. A non-load bearing wall carries no weight.

Lintel

A horizontal member for spanning an opening to support the structure above

Bond

The arrangement or pattern of laying bricks and blocks to spread the load through the wall, also for strength and appearance

Solid wall

Walls of a thickness of one brick and greater

Cavity wall

Walling built in two separate skins (usually of different materials) with a void held together by wall ties

ACTIVITY

What are the walls in the building you are sitting in made from? Why do you think these materials were chosen? What are the advantages or disadvantages of these materials?

Walls often have openings in them, eg doors and windows, which will weaken them if they are not constructed correctly. Openings require support (via a **lintel** or arch) across the top to give the wall support and **bond** it together.

EXTERNAL WALLING

External walls need to:

- keep the elements (wind and rain) out of the building
- look good
- fit into the surrounding landscape.

Several methods of construction are used for external walling. Common construction methods are:

- **solid wall**
- **cavity wall**
- timber framing.

SOLID WALL

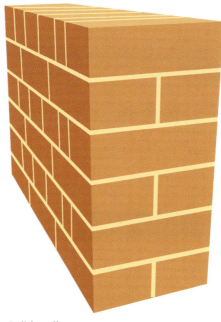

Solid wall

Many older traditional buildings have solid walls made from brick, block or stone, see the following table. Solid walls have the disadvantage of being more easily penetrated by damp. Older solid walls are often upgraded by having insulating and waterproofing layers applied to the outside of the wall.

Material used	Description
Bricks 	A very traditional building material made from fired clay, calcium-silicate or concrete. A standard sized brick is 215mm × 102.5mm × 65mm.
Blocks 	These are made of either concrete (crushed stone and cement) or a light-weight cement mixture. They are much bigger than a brick, and are available in various sizes. The most commonly used size is 440mm x 215mm x 100mm. Wider blocks are used for walls where a higher strength or improved sound insulation is required.
Stone 	A natural building material, which varies widely in use and appearance from area to area. Stone may be cut to a uniform size before use or used in its quarried state.
Mortar 	This is used between bricks, blocks and stones to bind them together and increase the strength of the wall. It is a mixture of soft sand and cement mixed with water and other additives if required, eg **plasticiser**, colouring or **lime**. It is important that the strength of the mortar is correct for the type of material that is being used to construct the wall. If the mortar has too much cement in the mix it will be so strong it will not allow movement in the walling due to settlement, and the bricks could crack resulting in the wall needing to be rebuilt. Mortars are mixed to a ratio of materials, eg 1:6. The first number is always the proportion of cement with the second being the proportion of sand. A typical mix ratio for masonry walling is 1:5.

Plasticiser

An additive that is used to make the mortar more pliable and easier to work with

Lime

A fine powdered material traditionally used in mortar

CAVITY WALL

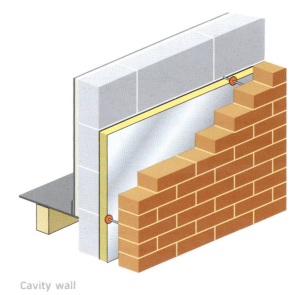

Cavity wall

ACTIVITY

State the minimum performance standards required to meet current building regulations.

ACTIVITY

Find out the current minimum width of cavity allowed.

Leaves

The two walls or skins that make up a cavity wall to comply with current building regulations

Building regulations

A series of documents that set out legal requirements for the standards of building work

The most common type of external walling used today is cavity wall construction.

Cavity walls are two masonry walls built side by side to form an inner and outer leaf (sometimes called skins). The **leaves** are held together with wall ties. These ties are made from rust and rot proof material and are built in as the walls are being constructed. The cavity is partially filled with insulation (typically fibreglass batts or polystyrene boards) as required by the **building regulations**. This reduces heat loss and saves energy.

The inner leaf usually carries any loads from the roof and floors down to the foundations and has a decorative finish on the inside, typically plaster which is either painted or papered. The outer leaf resists the elements and protects the inside of the building.

TIMBER FRAMING

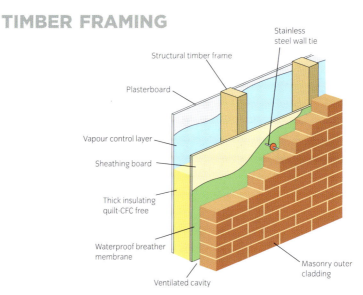

Timber frame wall

Timber framing is both a traditional and modern method of building. Traditional buildings using timber framing were made mostly from oak with various in-fills such as brick or plaster to form the walls. Modern timber frame homes are generally built from softwood and have an outer skin of masonry or are clad with timber or plaster to waterproof the structure. Oak framing, as a traditional building method, is becoming increasingly popular again.

Elizabethan oak frame

PREFABRICATED WALLS

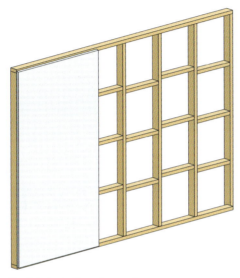

Prefabricated wall panel

There are a variety of prefabricated products available, generally made in a factory and then transported to site to be erected. These products enable quick and easy building. Often the **services** are pre-installed.

Services

Those provided by the utility companies, eg gas, electric and water

INTERNAL WALLING

Internal walling can be load or non-load bearing. Internal partitions divide large internal spaces into smaller rooms.

Internal partitions can be made from studwork or masonry. Studwork partitions consist of studs (which can be made from timber or metal) covered with a sheet material (usually plasterboard).

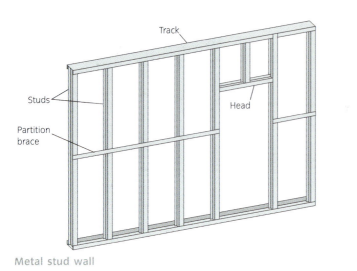

Metal stud wall

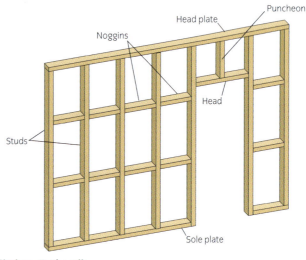

Timber stud wall

WALL FINISHES

External walls made from brick usually have no further finishes added while walls made from blocks are rendered. This is a covering of sand and cement mortar which is then finished with masonry paint.

Internal walls are most often plastered with a thin layer of gypsum plaster over plasterboard; this gives a very smooth hardwearing finish which is then usually finished with emulsion paint or papered coverings.

It is important to **size** new plaster to give a good base before applying further coverings of paint or paper coverings. This first coat of paint or paste is usually thinned down by 10% with clean water.

ROOFS

Roofs are designed to protect the structure below by keeping the weather out. As heat rises, the roof must be well insulated to prevent heat loss and improve the energy efficiency of the building.

TYPES OF ROOFS

Roofs come in a wide variety of designs as the following pictures show.

Pitched roof

Flat roof

INDUSTRY TIP

A flat roof has an incline of up to 10° while a pitched roof has an incline over 10°.

INDUSTRY TIP

Timber requires protection from the elements (rain, wind and sun) and this is done using timber coatings. Knotting is applied to prevent heat from the sun drawing resin out of knots in the timber. Primer is applied to give a good key to the paint or stain that is used to provide a finish. Paint also requires undercoat to be applied to give a good finish. Paint and stain can be water or solvent borne (water- or oil-based).

ROOF COMPONENTS

Roofs are commonly covered with slates or tiles. Slates are a natural product. Slate is a type of mineral that can be split into thin sheets. Artificial cement fibre slates are also available. Tiles can be made from clay or concrete.

Slate

Cement fibre slate

Roof tiles

A felt is laid below the roofing material to provide additional protection in case some water gets through the tiles.

Flashings are commonly made from lead and are used to provide waterproofing at joints where roofing materials meet walls and around chimneys.

Flashing providing waterproofing

Flashing around a chimney

SERVICES

Buildings contain services such as:

- water
- electricity
- gas supplies.

Additionally, waste such as sewage and water run-off have to be considered.

WATER

Water is brought into a building using pipes. Supply pipes used are usually made of plastic, with internal domestic plumbing being made from plastic or copper. Plumbing is installed using a variety of fittings including tees, elbows, and reducers. Bathrooms, kitchens and most heating systems require plumbing.

Copper pipe

Plastic pipe

Pipe fittings

Not only is water carried into a building, it is also taken away. Rainwater run-off is collected into gutters and taken away via downpipes and drains and returned to the ground or stored for later use.

Rainwater gutter flowing down pipes and into drain

SEWAGE

Sewage is taken away from the building via drains and is disposed of either into a sewer or into a septic tank/sewage treatment plant.

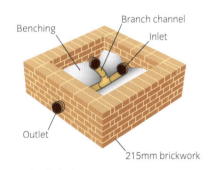

Benched drain

Septic tank

Sewage treatment plant

ELECTRICITY

Electricity is an important service provided to buildings. It powers lighting and heating. It is brought into a building through cables.

Electricity cables, switches and socket

Gas pipework to boiler

ACTIVITY

What services are being used in the building you are sitting in? How are they brought into the building?

GAS

Gas is brought into a building using pipes. Gas powers heating systems and provides fuel for cooking.

OTHER SERVICES

Other services that are installed include telephone systems and other data cables for broadband and entertainment systems.

SUSTAINABILITY

Our planet is a fixed size. Fossil fuels, eg oil and coal, which we take from the ground are not infinite, ie they will run out one day. However, the wind, the sun and the tides will always be there. These are sustainable sources of energy.

Building materials can be sustainable if they are chosen carefully. For example, the process of manufacturing concrete uses a lot of fuel and produces a lot of carbon dioxide (a gas that some say is damaging the climate).

On the other hand, trees absorb carbon dioxide as they grow, look nice and the timber they produce is an excellent building material. However, some timber is harvested from rainforests without thought for the surrounding environment or are harvested to such an extent that certain species are close to extinction. Managed forests where trees are replanted after harvesting provide a sustainable source of timber.

Here are some questions to consider regarding sustainability in construction.

MATERIALS

- How far have the materials been brought? Locally sourced materials do not have to be transported far, thus reducing fuel use.

- Are the materials sustainably sourced? Has the timber come from a managed forest or has it come from a rainforest with no regard to the environment?

- Have the materials been manufactured with the minimum of energy and waste?

DESIGN

Is there an alternative design that can be used that uses more sustainable materials? For example, a timber frame instead of concrete block or brick.

The table below shows some sustainable materials:

Material	Image
Straw bales	
Cob (soil)	
Timber	 Redwood Spruce Oak
Bamboo	

ENERGY EFFICIENCY

Energy is expensive and is only going to get more expensive. As the population increases more and more energy will be required. This needs to come from somewhere and its production can be damaging to the environment. The less power a building uses the better and if it can produce its own that is a bonus. Energy saving measures can save a lot of power consumption.

INSULATION

Light, air-filled materials tend to have better thermal insulation properties than heavy, dense materials. This means that heat cannot easily pass from one side to another and so if these materials are used in a building it will require less heating during the winter and will remain cooler during the summer.

The following drawing shows how much heat a typical home loses through different parts of the property. Better insulation will reduce the amount of heat lost.

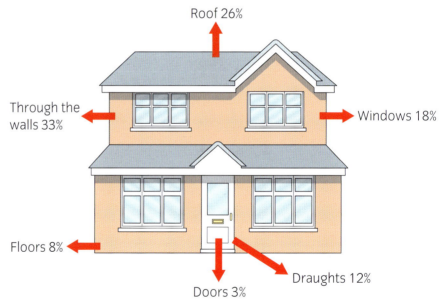

Sources of heat loss from a house

What insulation has been used in the building you are sitting in? Is the building energy efficient? Is it cold? Does it take a lot of heating? Take a look at 'Our House' and identify the insulation measures used there.

The table below shows some examples of insulation:

Type of insulation	Description
Blue jean and lambswool	Lambswool is a natural insulator. Blue jean insulation comes from recycled denim.
Fibreglass/Rockwool™	This is made from glass, often from old recycled bottles or mineral wool. It holds a lot of air within it and therefore is an excellent insulator. It is also cheap to produce. It does however use up a fair bit of room as it takes a good thickness to comply with building regulations. Similar products include plastic fibre insulation made from plastic bottles and lambswool.
PIR (polyisocyanurate)	This is a solid insulation with foil layers on the faces. It is lightweight, rigid and easy to cut and fit. It has excellent insulation properties. Polystyrene is similar to PIR. Although it is cheaper, its thermal properties are not as good.
Multifoil	A modern type of insulation made up of many layers of foil and thin insulation layers. These work by reflecting heat back into the building. Usually used in conjunction with other types of insulation.
Double glazing and draught proofing measures	The elimination of draughts and air flows reduces heat loss and improves efficiency.

MAKING BETTER USE OF EXISTING AND FREE ENERGY

SOLAR POWER

The sun always shines and during the day its light reaches the ground (even on cloudy days). This energy can be used. A simple use of this is to allow sunlight to enter a building. With a little thought in design, light can reach deep into a building via roof lights and light tunnels. This means that internal artificial lighting requirements are reduced, therefore saving energy.

Solar panels can generate hot water or electricity, and once the cost of installation has been covered the energy they produce is totally free.

Solar panel

A panel that absorbs sun rays to generate electricity or hot water

Solar panels

HEAT SOURCE AND RECOVERY

Humans give off a fair bit of energy as they go through a normal day (eg body heat, heat given off by hairdryers, cookers, refrigerators and other activities) and this can be conserved. Modern air-conditioning systems take the heat from stale air and put it into the fresh air coming in.

Heat can be taken from the ground and even the air outside.

WIND POWER

Wind power is becoming more widespread. However some people feel that wind turbines are damaging the visual environment as they spoil the appearance of the countryside. Individuals will have their own opinion on whether wind power is a good thing or not as there are many considerations to be taken into account.

Wind turbine

WATER POWER

Water is another source of power, whether that be hydro-electric (water from dams turning turbines) or wave power which is currently under development.

BIOMASS HEATING

Biomass heating (using wood and other non-fossil fuels) is also becoming more popular as these systems can heat water efficiently as well as heat rooms, and of course a well-insulated building does not require a lot of heating.

ENERGY EFFICIENT GOODS AND APPLIANCES

Energy efficient electrical goods (eg low energy light bulbs) and appliances (eg dishwashers, fridges and washing machines) which use a reduced amount of power and less water are available.

Case Study: Kayleigh

Kayleigh is to build a small single garage at the rear of a house. It must be big enough to accommodate an estate car and give enough room to allow the user to get out and walk around the car. The garage has two windows, an up-and-over door at the front and a flat roof. She has been asked to provide a plan of this garage for the client.

Draw this garage to a scale that will fit onto an A4 piece of paper. Include the window openings, the door, the thickness of the walls (which will be single block) and the piers.

Work through the following questions to check your learning.

1 What is the perimeter of this room?

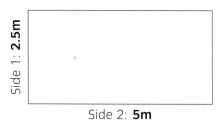

Side 1: **2.5m**

Side 2: **5m**

a   5m

b   7m

c   15m

d   17m

2 A message that is passed on by word of mouth rather than in writing is

a   Open to interpretation

b   Very accurate

c   Easy to understand if shouted

d   Easily remembered

3 What is a component drawing?

a   A plan of the whole building, floor by floor

b   A section through a part of the structure

c   An elevation of the walls

d   A detail in a room

4 What is the foundation type shown?

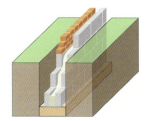

a   Strip

b   Pile

c   Raft

d   Pad

5 What is the foundation type shown?

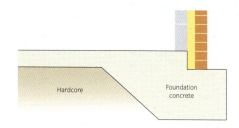

Hardcore   Foundation concrete

a   Strip

b   Pile

c   Raft

d   Pad

6 What is the component shown?

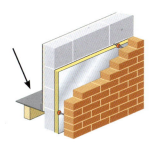

a   Damp proof membrane

b   Strip foundation

c   Damp proof course

d   Raft foundation

7 Which **one** of the following materials has the best thermal insulation properties?

a   Brick

b   Concrete

c   Glass

d   Polystyrene

8 Concrete sets because it contains

a   Aggregate

b   Sand

c   Hardcore

d   Cement

9 A flat roof has a pitch of less than

    a 8°

    b 10°

    c 12°

    d 15°

10 Load bearing walls transmit weight down to the

    a Foundations

    b Floors

    c Roof

    d Windows

Chapter 3
Unit 113: Maintain and use carpentry and joinery hand tools

Maintaining good quality tools will ensure safety, accuracy and help produce work of a high standard. Most of the tools we use today are very similar to those used over 100 years ago but we are also able to take full advantage of a newer range of tools available to us now. Whether using old or new tools we must remember that sharp tools don't cause accidents but blunt tools do, and using the right tools for the job will produce the fastest and most accurate results. Buying and using your tools will give you a great deal of pleasure. Your tools are also your shop window: their condition and how you look after them will often be reflected in your work.

By reading this chapter you will know how to:

1 Maintain, use and store hand tools.

2 Use measuring and marking tools.

3 Use hand saws.

4 Use hand-held planes.

5 Use woodworking chisels.

6 Use hand-held drills.

Heavy-duty handsaw

Risk assessments

A risk assessment is a careful examination of what, in your work, could cause harm to people, so that you can weigh up whether you have taken enough precautions or should do more to prevent harm

WORKING SAFELY WITH TOOLS

It is important to ensure the highest standards of safety when using, maintaining and storing hand tools. All training and workplaces are required to have a culture of safe working. This is developed from following good examples. In your training you will only be shown safe methods of working. By learning good working practices, you will promote this culture and recognise and report any unsafe methods you may see. You will be made aware of **risk assessments** that are in place and start writing your own (see Chapter 1, pages 5–6).

The following are general safety rules when using hand tools:

- Wear the correct type of PPE (personal protective equipment) for the task.

- Do not wear loose clothing, jewellery or trailing earphone wires that could get caught up in tools.

- Secure all work down to a bench or stools before working on it.

- Keep hands well away from cutting edges, such as saw teeth, drills and chisels.

- Never force the tool, it should always do the work.

- When handing a tool to someone, always offer them the handle end of the tool.

- Do not use tools for tasks for which they are not designed, as this often leads to accidents.

- Keep work areas clean and tidy (good housekeeping).

Hand tools covered in this chapter can be grouped into the following categories:

- measuring tools

- marking tools

- saws

- planes

- chisels

- drills.

MEASURING AND MARKING TOOLS

MEASURING TOOLS

For every job you work on, you will be working to a given size. This will generally have a **tolerance**, eg a door frame is required to be 900mm wide with a tolerance of +/– 2mm which means it is acceptable if it is made between 898mm and 902mm. It is important then that when either measuring or checking a measurement the rule you are using is accurate and you are using it to obtain the most accurate measurement possible.

STEEL TAPE MEASURES

Steel tape measures are very popular. They are available in a variety of lengths between 2–8m long. As the length of the tape increases so does the width of the tape to ensure stability over distance. The tape can be pulled out and locked at a specific position, then retracted when not in use for easy storage. They also have a belt clip for temporary storage in use.

The end of the tape has an adjustable hook (which allows for the thickness of the hook end) which ensures an accurate reading whether internal or external measurements are taken. However, the hook is a problem when measuring from a drawing, so to make sure accurate measurements are taken it is best practice to hold the tape on the first point at 100mm, then take away the 100mm to the measurement taken at the other end.

Using a tape measure from the 100mm point to take accurate measurements from a drawing

Tolerance

The deviation allowed from a specified size

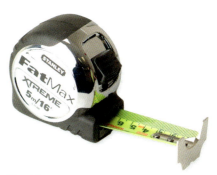

Steel tape measure

Adjustable hook on a tape measure

INDUSTRY TIP

Regular checks should be made to ensure that the tape is measuring accurately. If the adjustable hook gets damaged or bent they will give inaccurate readings.

ACTIVITY

Using a steel tape, measure the size of the nearest door. Write the size down in millimetres.

Steel rule

INDUSTRY TIP

150mm steel rules are the best tool to check the depth of a stub mortice.

INDUSTRY TIP

White chalk rubbed into the engraved edge of the steel rule will allow the measurement to be seen more clearly.

Four-fold rule

INDUSTRY TIP

An elastic band wound around a four-fold rule when not in use will prolong its life and keep it protected in storage.

Stub mortices

A mortice that does not pass through the full width of the timber

STEEL RULES

Steel rules are available in lengths from 150mm–2000mm. The short ones are very useful for fine accurate measurements and setting up gauges or specialist planes such as plough and rebate planes (discussed later in this chapter). The longer rules are heavy and this can be used to your advantage when **setting out** on paper or manufactured boards. They can also be used as a straight edge when drawing lines. When buying a rule choose a satin chrome finish: although more expensive they are much easier to read and are rust protected. Ordinary steel rules should be lightly oiled to prevent rusting in storage.

FOUR-FOLD RULES

Available in plastic and wood and normally 1000mm long, four-fold rules were traditionally used by carpenters and joiners but are losing favour to the steel tape measure. Because they are hinged, they are fragile and need careful use and storage. The better ones have brass inserts at their ends to ensure an accurate zero point is always available.

ADDITIONAL RULE USES

Apart from taking measurements, rules can be used to:

- Check the depth of **stub mortices**.

Checking the depth of a stub mortice

- Divide a board into any number of equal parts, eg: a board 127mm wide needs to be divided into five equal parts. 127 does not divide by five equally, so we place the rule diagonally across the board with zero on one edge and a larger measurement that easily divides by five on the other. In this case 150 has been chosen. 150mm ÷ 5 = 30mm. Mark multiples of 30 from the rule on to the timber. These lines when 'ruled off' parallel to the edge will give an exact division of the board into five.

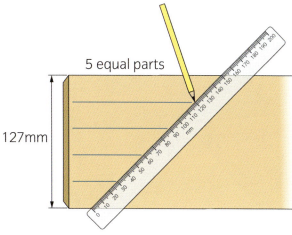

5 equal parts

127mm

Dividing a board into equal parts

- Rule off. In other words to draw lines parallel to the timber's edge using the rule and your hand as a stock.

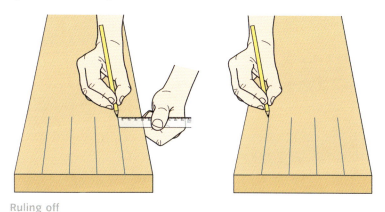

Ruling off

TOOLS THAT MARK ANGLES

The most common angle required to be drawn is a **right angle**. Another angle frequently required is 45°, most commonly required for **mitre joints**. Occasionally other non-standard angles are required and we need to be able to accurately mark these as well.

TRY SQUARES

Try squares are used wherever a right angle is required to be marked or tested for accuracy. They consist of a hardwood stock and a steel blade. The blade is secured to the stock by pins. They are available in sizes from 100mm–300mm. While still commonly used, they are being overtaken by the more versatile combination square.

Right angle

A right angle contains 90°

Mitre joint

An internal or external junction of a moulding

Try square

Face side and edge

The best two faces of the timber

For marking out to be accurate, a square line must be marked around a piece of timber with the stock only against the **face side and edge**.

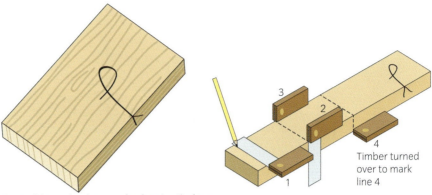

Face side and edge marked onto timber

Timber turned over to mark line 4

ACTIVITY

Check any square using this process.

A try square can be checked for accuracy as follows:

1 Place the stock of the try square against a straight piece of timber and strike a pencil line on the timber.

2 Rotate the stock of the try square through 180° and check against the first line drawn.

3 If it is in the same place the try square is accurate and can be used to check for accurate right angles.

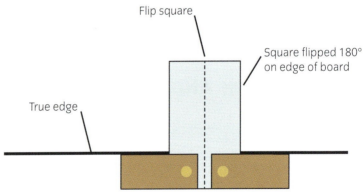

Flip square

Square flipped 180° on edge of board

True edge

Testing a square

SET MITRES

The set mitre is sometimes incorrectly known as a mitre square (how can a mitre be square?), and can only be used to mark 45° angles. It is rarely purchased nowadays as, again, the combination square can do the job of this and a try square.

Set mitre

COMBINATION SQUARES

The combination square is all metal in construction. The blade on a combination square can slide through the stock and be locked into any position along the length of the blade. As the name suggests they can be used for a number of purposes but they are primarily used for marking and testing for square. The shape of the stock is also designed to give you an angle of 45°, so can replace the set mitre. Additionally it can be used for ruling off, checking the depth of a stub mortice and measuring. The better quality ones also have a spirit level (more ornament than useful) and a scribing pin (for marking metal) which is located in the bottom of the stock.

Combination square

SLIDING BEVELS

The sliding bevel is constructed of a hardwood stock, a sliding blade and a locking screw. This tool is used to obtain and test *any* angle including 45 or 90°. The image below shows a sliding bevel being set up to a pitch (angle) of 1:6 for the marking of dovetails.

Sliding bevel

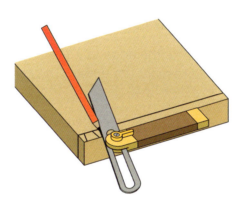

Setting up a sliding bevel to a pitch of 1:6

Sliding bevel used to mark out dovetail joints

INDUSTRY TIP

To accurately mark a line around a piece of timber, the stock of the square must only be placed against the face side and edge as shown.

TOOLS THAT PRODUCE LINES PARALLEL TO THE EDGE OF TIMBER

GAUGES

Often found to be the hardest tool to use early in training, gauges are used to mark out accurate lines parallel to the edge of a piece of timber. If not used correctly the gauge pin will follow the direction of the grain rather than run parallel to the edge of the timber. Most joints are marked out using one of the following gauges.

Marking gauge

Marking gauge

The marking gauge consists of a hardwood stock and stem, a pin and thumb screw. Again, better quality tools have brass inlays to reduce wear on contact surfaces. The distance between the stock and pin can either be set using a rule or, if being used to find the centre of a piece of timber, use the following method:

1 Set the distance between the pin and stock to approximately half the timber's thickness.

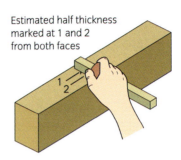

2 Lightly tighten the thumbscrew.

3 Holding the stock tightly against the edge of the timber, roll the pin lightly to mark a line.

4 Reverse the edge that the stock is against to the opposite side and re-mark the timber. The centre will be between these two marks.

5 Adjust the gauge by either tapping the bottom of the gauge to increase the distance, or the top of the gauge to decrease the distance.

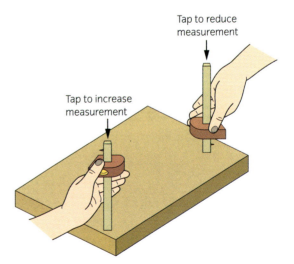

INDUSTRY TIP

Gauges can be pushed or pulled (with the gauge pins trailing the movement). Use it whichever way you find easiest.

ACTIVITY

Carry out the process to find the centre of a piece of timber.

6 Repeat stages 4 and 5 until the pin mark is in the same place, and then fully tighten the thumbscrew. You have found the centre of the timber.

Mortice gauge

The mortice gauge is similar to the marking gauge but it has two pins rather than one. The stem has a mechanism to allow the adjustment of the gap between the two pins. The construction of the gauge is generally more robust than the marking gauge and is normally only carried by a bench joiner these days.

Primarily, the mortice gauge is used for marking out mortice and tenon or bridle joints.

Combination gauge

A combination gauge is very similar to a mortice gauge but with an additional pin on the opposite side of the stem so that it can be used as a marking gauge.

Cutting gauge

A cutting gauge is the same as a marking gauge except instead of a pin it has a cutting knife and a wedge to hold it in place. It is used if a parallel gauge line is required across the grain, for example when marking out dovetails. The knife severs the fibres of the grain leaving a clean cut. If a marking gauge was used it would pull the fibres of the grain up leaving an unacceptable finish.

Mortice gauge

Combination gauge

Cutting gauge

Tooling

The part of the tool that cuts the timber, ie circular saw blade

HAND-HELD SAWS

With the increasing use of power tools, saws are being used less and less, particularly on site. Hand-held circular saws and jigsaws have **tooling** that replicates the cutting action of traditional saws. It is important then to ensure that we know what type of saw is best selected for the task in hand. The teeth on saws vary considerably in size. We normally measure this by how many teeth (the part that does the cutting) there are per 25mm. The general rule is, the thicker the timber being cut, the larger the saw (and therefore the teeth on the saw) required.

Timber can be cut along the grain (called ripping) and across the grain (crosscutting). To ensure the most efficient cut, a different tooth design is required for each operation.

HOW THE SAW WORKS

The cutting action of rip saw teeth can be best described as a series of chisels each cutting its own groove. The cutting action of crosscut teeth are shaped to act as a series of scored knife cuts, so their edges sever the fibres of the timber and produce a clean cut.

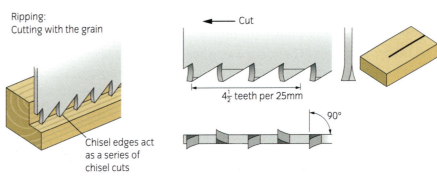

Ripping:
Cutting with the grain

Cut

4½ teeth per 25mm

90°

Chisel edges act
as a series of
chisel cuts

Cutting actions of a rip saw

Set

Each alternate tooth is bent in opposite directions

Binding

When a saw jams during a cut

The teeth of traditional saws are **set**. The increased width of cut created by the set is called the kerf. This clearance stops the saw from **binding** in the cut and makes for an efficient cut. The gaps between the teeth are called the gullets. These store the waste timber (sawdust) until the teeth are exposed from the timber they are cutting.

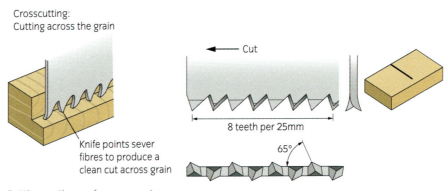

Crosscutting:
Cutting across the grain

Cut

8 teeth per 25mm

65°

Knife points sever
fibres to produce a
clean cut across grain

Cutting actions of a crosscut saw

ACTIVITY

Search on the internet to find the type of steel that hand saws are made from.

Both these traditional actions cut on the push stroke. Additionally, we are now more commonly using Japanese-style saws which are much more efficient at cutting timber. These have a more complex design of tooth and cut on the pull stroke.

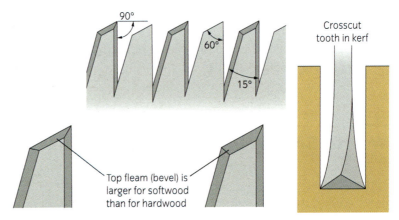

90°

60°

15°

Crosscut
tooth in kerf

Top fleam (bevel) is
larger for softwood
than for hardwood

Teeth design of Japanese saws

TYPES OF SAW

RIP SAW

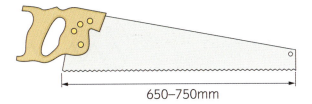

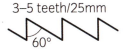

3–5 teeth/25mm

60°

650–750mm

Rip saw

The rip saw is now very rarely used as carpenters and joiners are able to use powered alternatives and the nature of our work has changed so they are not required. We do, however, need to know their purpose and understand why their teeth are sized and shaped as they are as this is the basis of how the teeth on tooling used on powered hand tools are designed.

Rip saws are the largest of the hand saws, between 650mm and 750mm long with between 3 and 5 teeth per 25mm. They are designed for cutting down the length of the grain (commonly called ripping).

Rip saws should never be used to cut across the grain as the cutting action would, if managed, produce a very poor quality of cut or **spelching**. The saw may also jump in the cut and possibly cause an accident.

Rip saws in use
When sawing, the operator should be using the full length of the saw for the most efficient cut. A short stroke simply tires the operator and creates uneven wear on the saw, blunting the centre section and leaving both ends untouched. The operator should also let the saw do the work rather than force it through the timber.

Spelching

Damage at the end of a cut where the unsupported grain breaks away

A rip saw in use

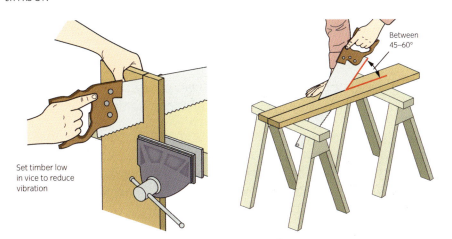

Set timber low in vice to reduce vibration

Between 45–60°

Sharpening angles for rip and crosscutting saw teeth

Crosscut saw

CROSSCUT SAWS

Still very commonly used, traditional crosscut saws are 600–650mm long with 6–8 teeth per 25mm. They are used to cut large sections of timber such as floor joists.

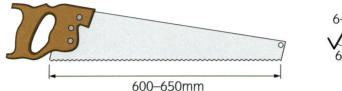

6–8 teeth/25mm

60°

600–650mm

Crosscut saw

A smaller version of the crosscut saw is called a panel saw. It is between 500–550mm long with smaller teeth, and has about 8–10 teeth per 25mm. It is used for finer work such as cutting smaller sections, for example floor boards, skirting and **manufactured boards**.

Maunfactured boards

Man-made panel products such as plywood and Medium Density Fibreboard (MDF)

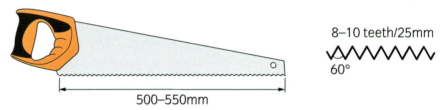

8–10 teeth/25mm

60°

500–550mm

Panel saw

Hard point

Where the teeth are specially hardened during manufacture

We tend to use **hard point** panel saws almost exclusively as power tools are now usually used for heavier cuts. These disposable saws are particularly popular. They hold their sharp edge for a much longer period, but once blunt they cannot be re-sharpened and have to be thrown away.

Crosscut saws in use

All cuts should be made against a pre-marked line. The saw cut is started by pulling back two or three times against the thumb. The thumb is used as a guide for the accurate positioning of the saw and it also helps prevent the saw from jumping at the beginning of the cut as it improves control. When the cut is established on the pull stroke, the thumb can be removed and the operator can start pushing the saw following the pencil line. Crosscut saws can be used to cut down or across the grain (although less efficiently at cutting down the grain).

When cutting manufactured boards it is important to ensure that the boards are well supported to avoid accidents. The size of boards makes cutting difficult as stretching across the board is often required. Assistance may also be needed to support the offcut. The angle of the saw should be kept lower to ensure a clean cut on the underside face.

INDUSTRY TIP

Keep the saw cut on the waste side of the line. This will ensure the correct size is achieved.

This sequence shows you how to use a crosscut saw.

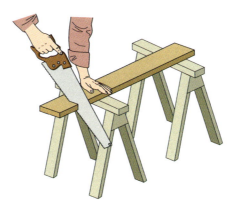

STEP 1 Begin your cuts by drawing back (pulling the saw back) against the thumb.

STEP 2 Cut the timber to length, supported by stools.

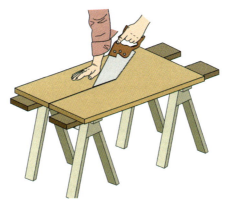

STEP 3 Support the offcut to prevent damage to the board.

STEP 4 Cutting manufactured boards requires support across stools and additional assistance.

BACKED SAWS

Backed saws, such as tenon and dovetail saws, have either a strip of steel or brass that runs along the top edge of the saw. Its function is to keep the blade taut, straight and to add weight to the saw. The best saws have a heavyweight brass back. Backed saws are used for the accurate cutting of joints and can be used to cut with and across the grain.

Architrave

A moulded section that is fixed around a door lining that covers the gap between the lining and the wall

Tenon saw

The tenon saw is used for both the cheek and shoulder cuts of a variety of mortice and tenon type joints at the bench and is also used for cutting **architraves** on site. It has 12–14 teeth per 25mm, available in both traditional and hard point, in lengths of 300–350mm long.

Tenon saw

Using a tenon saw to cut timber to length in the vice with the aid of a bench hook

Dovetail saw

Used mainly for cutting dovetail joints and other light work, dovetail saws have very small teeth, 16–22 per 25mm, and are generally always 200mm long. They can be easily damaged if used on work that is too large because their blade is very thin.

Dovetail saw

OTHER SAWS

The following are more saws you may come across.

MITRE SAW

Often used in early training, the mitre saw can be set to 45°, 90° and any other angle inbetween. Its frame ensures it always cuts square through the thickness of the timber. Its size coupled with the skill you will gain makes it redundant as your training progresses. As its name suggests, the mitre saw is commonly used for cutting the mitres required for moulding intersections. Traditionally a mitre cut would be made then the moulding cut to size using a tenon saw with this aid.

Mitre saw

Mitre cut/box

COPING SAW

The coping saw consists of a narrow blade, held in tension within a frame. A handle is attached to the frame for control. It is commonly used to remove the waste material when cutting dovetail joints, to cut scribed moulding profiles and for general curved cutting. Its use is limited by the size of its frame.

Coping saw

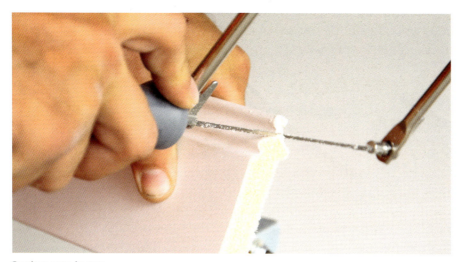

Coping saw in use

PAD SAW

The pad saw has been largely replaced by a portable powered jigsaw. It has a narrow blade which allows curved cuts to be made. When not in use the blade can be stored in its handle for protection and safety when being carried. Some utility knifes have blades that can be inserted and adapted to be used as a pad saw. As the blade can only be used on the push stroke they are prone to bending.

Utility knife type pad saw

Traditional pad saw

Fleam

The angle of the faces of the teeth at a right angle to the face of the saw

JAPANESE SAWS

For bench work, Japanese saws cannot be surpassed. They have extremely efficient cutting **fleam** teeth, which cut primarily on the pull but also the push stroke. The pull cut keeps the blade tense and maintains the line of cut. The teeth cut very quickly so you need to have good control to ensure you do not cut too far.

Japanese saws: top to bottom, Ryoba, Katuba and Dozuki

MAINTENANCE OF SAWS

Very few carpenters and joiners sharpen their own saws. When saws become blunt, they can be taken to your local tool shop where they will be sent to a saw doctor to be sharpened on a machine. This can be quite expensive which is why disposable saws are so popular. There is no substitute for buying good quality saws as these will last your working life. Below are photos of the equipment you will need for sharpening your saws.

Saw horse

Mill file

Saw file

Saw set

First the saw is set up in a saw horse, then the process for sharpening saws by hand is as follows:

1 *Topping.* The process of levelling the high (unworn) points to the lower (worn) points of the saw. This is achieved by lightly running a mill file along the top edge of the saw until the teeth are level.

2 *Shaping.* Bringing the teeth back to their correct shape and size using a saw file.

3 *Setting.* Use a saw set to bend each tooth alternately. This creates a kerf of the saw cut.

4 *Sharpening.* Filing to correct angle depending on cutting action required using a triangular saw file.

At the end of this process a sharpening stone is lightly run against both sides of the saw's teeth. This is called side dressing and it levels the projection of the set, improving the smoothness of the cut in use.

ACTIVITY

Go to a local tool shop or search the internet to find out how much it costs for:
1 a good quality hand saw
2 a panel saw
3 a good quality disposable saw.

STORAGE AND PROTECTION OF SAWS

Carpenters and joiners no longer make or carry traditional tool boxes, which stored tools and kept them safe. Various tool storage carts can be purchased, but few will take a hand saw.

To make sure saws are working at their best and are safe to carry, the teeth must be protected. When saws are bought they come with a card sleeve/sheath. These should be used as long as possible. They don't last forever and should be replaced with either plastic saw guards or purpose-made sheaths.

Plastic saw guard and card cover

Silicone spray

Saw guards are inexpensive but do not protect the blade from rusting if it is exposed to the elements. Purpose-made sheaths are available and have silicone treated materials or granules which protect the steel from rusting. In any event, saw blades should be lightly sprayed with silicone spray which protects the blades and makes the cutting action more efficient by reducing the effect of the **friction.**

Friction

The rubbing of the saw in the cut

OUR HOUSE

Look in 'Our House' and think about which saws you would choose for cutting the joists.

HAND-HELD PLANES

There is a considerable number of different types of hand-plane. All are designed to remove shavings of timber to produce flat or shaped surfaces, or to create a profile such as a groove or rebate. All planes require a great deal of skill to set them up and make them work well to produce good results. Hand-planes fall into two main categories:

- bench planes
- specialist planes.

BENCH PLANES

While they do not have to be used at the bench, most bench planes are. They all have a similar appearance but their length determines their use. The following image shows the different parts of a plane. All bench planes are set and adjusted in the same way.

ACTIVITY

Research the cost of two commonly used bench planes and state a use for each of them.

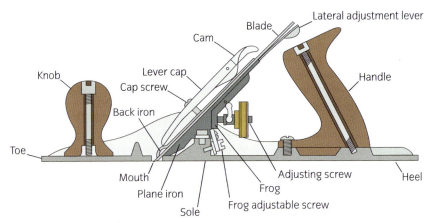

Different parts of a bench plane

TRY PLANE

The longest of the bench planes, the try plane was traditionally used to **true** timber prior to jointing in its width. This is not often done today as the use of wide, solid timber is rare. There are three long planes available:

- the fore plane – 500mm long

- the try plane – 550mm long

- the jointer – 600mm long.

The length of the sole (bottom part of the plane) of these planes means they will always take off the high points before reaching any hollows in the edge of the timber. The following image shows how a try plane is used to true the edge of a timber board.

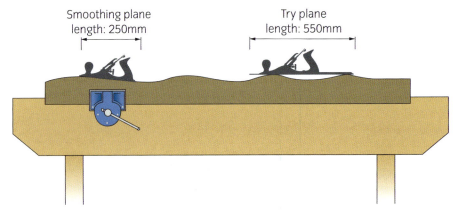

Short planes will only ride the hollows in wood, longer planes will remove 'high spots' to produce a straight edge

Try plane

True

To be completely straight and square edged

JACK PLANE

As the name suggests, the jack plane is a general purpose plane. It is commonly used for the preparation of sawn timber. It is 375–400mm long and is available in two widths of plane (cutting) iron, 50mm (number 5) and 63mm (number 5½). On site its most common use is to **shoot** doors into a lining/frame.

Jack plane

SMOOTHING PLANE

Probably the most commonly used plane, the smoothing plane is the shortest and lightest and is used for:

- fine planing of surfaces prior to assembly (cleaning off any pencil or machine marks)

- the flushing of shoulders and surface preparation after assembly.

Being only 250mm long, it is not suitable for truing long timber as it would simply follow its original shape. Again, available with two common iron widths, 50mm and 63mm.

Shoot

To plane to fit

Smoothing plane

INDUSTRY TIP

Flatten the back iron on a sharpening stone to ensure a close fit to the cutting iron. This will prevent shavings from getting trapped between the two.

MAINTAINING STANDARDS FOR BENCH PLANES

Regardless of the bench plane type, there are certain basic standards that must be met to ensure their correct use and to achieve a good finish.

1 The frog of the plane iron is set to achieve a mouth width suitable for the task in hand. A large mouth is required for preparing timber and planing wet or damp timber. A small gap is required for fine smoothing and end grain planing. The frog can be adjusted as shown in the image below.

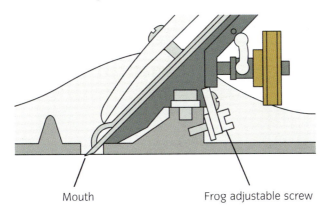

Mouth   Frog adjustable screw

Adjust the frog of a plane to either increase or decrease the size of the mouth

2 The plane iron is ground to the correct shape and angle for the work being carried out.

3 The back iron should be set back from the cutting edge, between 0.5mm for smoothing and end grain work and 1.5mm for preparing timber and wet or damp timber. The purpose of the back iron is to stiffen the cutting iron. This helps to minimise vibration in use and to break the shaving, causing it to curl and clear the mouth to prevent clogging.

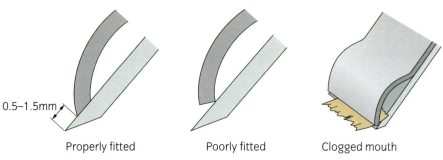

0.5–1.5mm

| Properly fitted | Poorly fitted | Clogged mouth |

Properly and poorly fitted back irons

4 The plane iron should be set so the cutting iron just projects through the mouth and is adjusted parallel to the sole. First, look along the bottom of the plane then adjust the lateral lever to bring the cutting edge to parallel with the sole. When parallel, wind the iron back so that just a hair's thickness is showing.

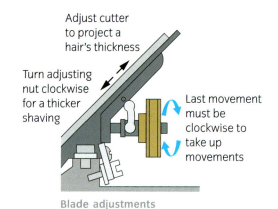

Adjust cutter to project a hair's thickness

Turn adjusting nut clockwise for a thicker shaving

Last movement must be clockwise to take up movements

Blade adjustments

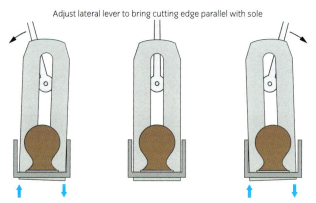

Adjust lateral lever to bring cutting edge parallel with sole

Looking along the bottom of the plane and adjusting the iron.

5 The direction of the planing should lay the grain down, giving a clean finish rather than lifting the grain (tearing), which gives a very poor finish.

Planing direction (torn against the grain)

6 When in use, pressure should only be placed over the part of the plane in contact with the timber. If not, the weight of the plane will tend to produce a bowed face.

Now the plane is set correctly it will be ready to use.

PLANES IN USE

PLANING TIMBER TO SIZE FROM SAWN STOCK

This is probably one of the first activities you will carry out in training. It will also probably be the last time you will do it. Nevertheless, it is essential to know how to, and be able to carry out this process as so much is dependent on it being done correctly.

Sawn timber is available in standard widths and thicknesses. British standards give 3mm per face planing allowance when preparing

INDUSTRY TIP

'A little bit of wax works wonders'. Wax the sole of the plane with a candle to reduce the friction and make working easier.

timber. So if we start off with sawn timber 100mm wide and 38mm thick (known as '100x38'), we would be allowed to plane it down to 94mm x 32mm. The 3mm planing allowance allows us to lose all sawn marks, flatten the faces and bring to parallel and square.

Planing process

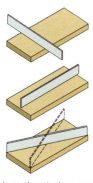

STEP 1 Look at the timber and select the face most free from **defects** and place this face up in the vice. A jack plane is used to remove saw marks and produce a clean flat face. This is checked with the edge of the plane, or straight edge, as shown.

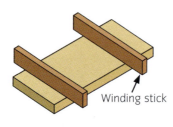

Winding stick

STEP 2 Use **winding sticks** to ensure the timber face is not twisted. Plane diagonally to remove the high spots if required followed by a few more shavings taken parallel to the edge of the timber.

STEP 3 Once this is complete use a standard reference mark to show the face (this should point to the best edge and is known as the face mark).

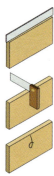

STEP 4 Place the timber in the vice with the planed edge facing out and the best edge pointing up. Plane the edge until it is straight and square to the face, which is checked using a straight edge and try square.

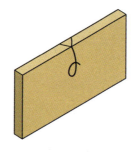

STEP 5 When correct, apply the remaining reference mark (known as the face edge mark).

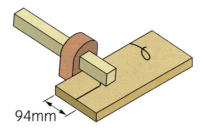

94mm

STEP 6 Next, the timber is gauged to width (94mm) using a marking gauge. Place timber in vice and plane down to the gauge line. Be careful to check the timber on all faces to ensure that the gauge line is being kept to.

Defects

Knots and splits etc

Winding sticks

Lightweight strips of parallel timber. Their length exaggerates the width of the timber to highlight any twisting

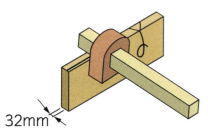

32mm

STEP 7 Lastly, gauge timber to thickness (32mm), place in vice and plane down to gauge line.

ACTIVITY

Now try the above process, the timber size doesn't have to be the same.

PLANING END GRAIN

End grain is more difficult to plane than long grain. Generally a lot more pressure is required to avoid the plane juddering along and giving a poor finish. To avoid break out we need to adopt one of the following methods of working:

- Plane from both ends to the centre

- Remove the far end corner so there is nothing to breakout

- Using a shooing board that supports the far edge and prevents break out.

FLUSHING SHOULDERS OF ASSEMBLED FRAMES

This is where a smoothing plane is used to level off the shoulders of joints. The edge of the plane is used to find any high points which are then gradually planed down to produce a flat surface across the joints. Be very careful not to **dub off** the **stiles** of the frame or to allow the back edge of the plane to drop as this will damage the inside edge of the frame. When the frame is flat, a couple of shavings are taken through the **rails** and finally through the stiles. Take care to ensure you are laying the grain down when choosing your planing direction.

Dub off
Over-plane the outside edge, leaving it thin

Stiles
The vertical parts of a frame

Rails
The horizontal parts of a frame

Flushing shoulders of assembled frames

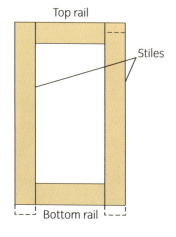

CUTTER ARRANGEMENT IN BENCH PLANES.

The cutter arrangement is the same for all bench planes with the grinding angle sitting down into the frog of the plane as shown on page 128.

SPECIALIST PLANES

Specialist planes are expensive to purchase and not often used. Bench joiners would generally have greater call on them than the site carpenter, perhaps with the exception of a block plane. Most of these planes have no back iron (compass plane excepted, which is the same as a bench plane). The cutter arrangement will also change depending on the plane.

BLOCK PLANE

Block plane

The block plane is a very convenient plane which can be used one handed. Better models have an adjustable mouth which can be set small for fine work or wide for coarser work. It has a much lower frog angle improving the finish of the cut. Typical uses would be trimming down edgings to plastic laminate or solid surface work surfaces and end grain work, such as adjusting the mitres when fitting architraves on site.

Mitre adjustment using a block plane one handed

REBATE PLANE

Rebate plane

As the name suggests, the rebate plane's prime use is to run rebates. The plane has a fence to control the width of the rebate (the better models have two arms) and a depth stop to set the depth of the rebate. In using this plane, you would start at the front of the timber and gradually work your way back. This will ensure you are laying the grain down and not tearing it up.

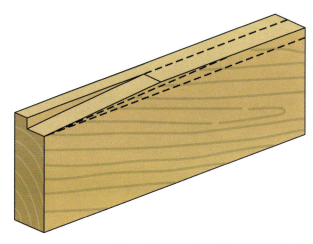

Start at the front of the timber and work your way back

A common fault when setting this plane up is not having the blade set just proud (sticking out) of the edge of the plane. If it isn't, a stepped rebate will be formed.

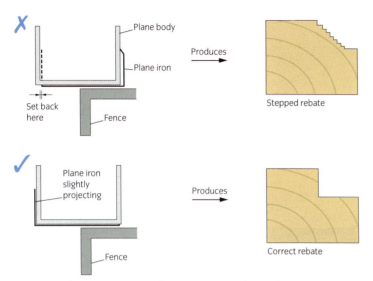

Correct and incorrect way of setting up a rebate plane

SHOULDER PLANE

The shoulder plane was traditionally used for truing up wide shoulders of tenons when they were cut by hand. Can also be used to correct or increase the size of rebates.

Shoulder plane

BULLNOSE PLANE

The bullnose plane is a short plane with the iron set within 10mm of its front end and is often incorrectly used as a shoulder plane. It is only suitable for working into corners in a stopped rebate or to adjust a rebate in an assembled frame. To be used correctly a pocket needs to be chopped out using a chisel. The bullnose plane can then be run into it.

Bullnose plane

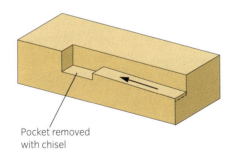

A pocket in a stopped rebate is needed when using a bullnose plane to work into corners

SIDE REBATE PLANE

The side rebate plane is used mainly to increase the width of a groove when fitting a panel.

Side rebate planes

Plough plane

Letting in

Fitting flush with the surface

PLOUGH PLANE

A plough plane is used to plough grooves along the edge of timber. A variety of plough irons are supplied to plough grooves between 3mm and 16mm wide. It is used in the same way as a rebate plane, starting at the front and working your way back.

ROUTER PLANE

Router planes are available in two sizes, the smaller version is called a thumb router. The larger version is used to level the bottom of housings to bring them to a regular depth. The smaller version can be very useful when **letting in** ironmongery, eg the face plate of a lock.

Router plane

Smaller router plane (thumb router)

Convex

Curved outwards

Concave

Curved inwards

COMPASS PLANE

The compass plane is used for planing curved joinery items accurately to shape. The sole of this plane is flexible and can be adjusted to **convex** and **concave** shapes. It is the only specialist plane that has a back iron. The nature of curved work means that it is almost impossible to plane with the grain throughout the length of the timber being planed. To avoid tearing out, the direction of planing has to be changed frequently.

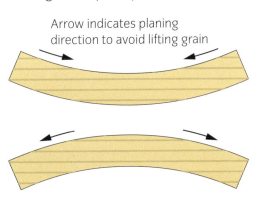

Arrow indicates planing direction to avoid lifting grain

Compass plane in use

Planing directions on curved work to avoid tearing up the grain

SPOKESHAVE

Spokeshaves are available with a flat bottom for planing convex surfaces and a curved bottom for planing concave surfaces. Like the compass plane, they are used to accurately shape sawn curves. Again, the direction of planing will have to be chosen carefully to avoid tearing out the grain.

Spokeshaves

CUTTER ARRANGEMENT IN SPECIALIST PLANES

You would use the following grinding angle up for block, shoulder, bullnose, side rebate plane and router planes.

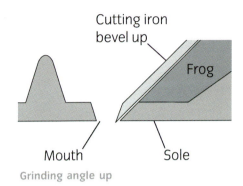

Grinding angle up

For spokeshaves, compass, rebate and plough planes, you would use the grinding angle down.

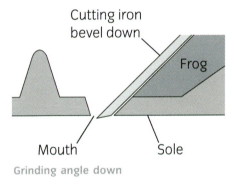

Grinding angle down

STORAGE OF PLANES

When being stored or transported, the cutting irons should be wound back into the body of the plane. They can be wrapped in cotton rags and sprayed with a rust protection solution. Often purpose-made boxes can be made to protect specialist planes such as shoulder planes. Before re-use, planes should be cleaned of any oil or rust protection to prevent staining of timber.

All planes require a great deal of skill to achieve good results. Don't be put off, this will come quickly with practice. You always know when a plane is working correctly as it will make a whistling sound when being used, listen out for it.

For maintenance of plane irons, see page 135 for grinding and page 139 for sharpening.

WOODWORKING CHISELS

There is a range of woodworking chisels available, most have a specific use as shown in the following table. There is however one golden rule: *keep both hands behind the cutting edge.* Stick to this rule and you will never suffer from a chisel cut. Always purchase the best quality chisels you can afford as these are not disposable items and should last you most of your working life.

The following table shows the characteristics and uses of common chisels:

INDUSTRY TIP

The 'tang' is the metal part of a chisel which fits into the handle.

Chisel type	Use	Section view	Common sizes available in mm
Mortice	A very strong chisel designed to withstand the heavy mallet blows required when chopping mortices and other heavy work.		6, 8,10 and 13.
Registered	Used for general heavy work on site and or wide mortices. The ferrule on the end of the handle protects the wooden handle when used with a claw hammer. Rarely used.		13, 16, 19 and 25.
Firmer	A general purpose chisel traditionally used on site as it is more robust than a bevel edged chisel. They are losing their popularity in favour of the bevel edged chisel.		6, 9, 13, 16, 19, 25, 32 and 38.
Bevel edged	Traditionally a bench joiner's chisel used for general work, paring and recessing etc. It should be the chisel of choice when chopping out the sockets for dovetails as the bevelled edge will not damage the sides of the dovetails (see woodworking joints).		As above. Although a 50mm wide chisel is available from certain manufacturers.
Skew ground bevel edged	A normal bevel edged chisel which has been ground at an angle (skew) specifically for cleaning out the sockets of lapped dovetails. Normally a pair would be ground (see joints).		Normally 12 and 16.

Chisel type	Use	Section view	Common sizes available in mm
Paring	A chisel with a very long blade, available in bevel edged or firmer varieties. It is used to remove the waste of long housings when produced by hand, eg **stair string** housings. This chisel should only be used with hand pressure and not used with a mallet as it is fragile due to the length of the blade.	or	19, 25, 32 and 38.
Butt	Could be either a well-worn bevel edged or firmer chisel with the handle cut down. Very handy if having to work in confined spaces.	or	To suit purpose required.
Scribing gouge Internally ground (cannalled) Blade Blade ground on inside	Used by both carpenters and joiners. They come in a variety of sizes and radii. They are primarily used when scribing (cutting out to fit over a mouldings profile) joints.		Variety of radii.
Firmer gouge Externally ground Blade Blade ground on outside	Mainly used by wood carvers, rarely used by carpenters and joiners.		Variety of radii.

Stair string

An inclined board each side of the stair to carry the treads

ACTIVITY

Use the internet to research three additional specialist chisels not listed in the table above.

Mallet

OUR HOUSE

Looking at 'Our House', think about where a chisel would be most useful in building the house. What sort of chisel would you use and why?

CHISELS IN USE

MALLET

We can't look at chisels without talking about the mallet. Traditionally made by carpenters and joiners, mallets are generally now purchased. They are most commonly made from beech (a European hardwood) and are available in a variety of sizes; a medium size is the most popular.

Mallets should be held at their far end (and not under the head, choking it to death) beacuse this is the most efficient use. All chisels, with the exception of registered and paring, should be driven with a mallet. While plastic handled chisels can be used with a hammer (and often are on site) it is better practice to use the mallet.

INDUSTRY TIP

When designing a mortice and tenon joint choose a width that a mortice chisel is available in.

CHOPPING RECESSES

A recess is a space chopped out of timber to receive other timber (joint) or piece of ironmongery. A typical example would be the recess for a hinge as shown in the image below.

An operative putting a hinge in a hinge recess

The following step by step guide shows how to chop a recess out of timber:

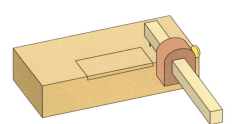

STEP 1 Mark out the recess to be cut using a gauge.

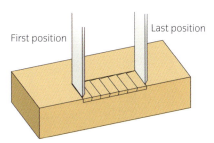

First position   Last position

STEP 2 Position chisel 3mm inside the recess end with the bevel towards you and start chopping out the recess using a walking method.

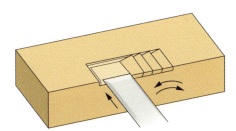

STEP 3 Remove the timber which has been chopped by paring out the waste down to the gauge line. Removing the bulk of the waste removes the resistance for the next operation.

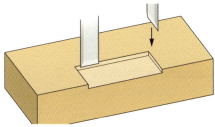

STEP 4 Carefully pare back to the gauge lines to ensure a tight fit. When paring, try to adapt a **shear** cut as this will give a better finish.

Shear

A cut using a slicing action

Common faults

Common faults would be taking down the recess below the gauge line or not positioning the chisel in the gauge line resulting in a recess not parallel to the face of the timber.

CHOPPING MORTICES

Mortices form the female part of a mortice and tenon joint, see Chapter 5, page 208. The mortice must be accurately chopped to ensure a tight fit with the tenon and a long lasting joint.

The following step by step guide shows how to use a mallet and chisel:

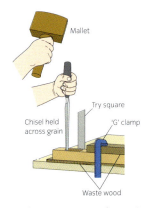

STEP 1 Set up your equipment as shown. Position timber over the bench leg and secure firmly. Position chisel between two gauge lines. Align chisel parallel and square to timber (do not allow the chisel to twist).

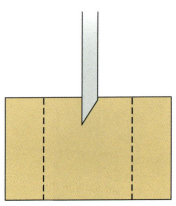

STEP 2 Start with the chisel in the centre of the mortice vertically between the two gauge lines.

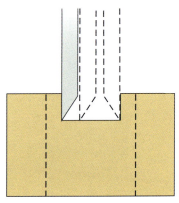

STEP 3 Work to either end of the mortice, ensuring the flat of the chisel faces the end of the mortice. Stop about 6mm from each end.

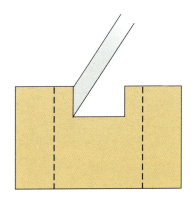

STEP 4 Turn the chisel bevel down and lever out the waste to clear the mortice.

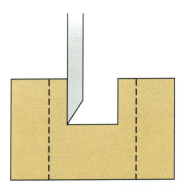

STEP 5 Chop to just over half the depth using the chisel and mallet.

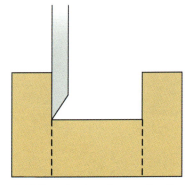

STEP 6 Chop back to the mortice lines. Turn the timber over and repeat from the other side.

> **INDUSTRY TIP**
>
> Leaving 6mm from each end when chopping mortices will prevent damage to the timber when levering out the waste.

Common faults

Common faults are wide mortices, caused by twisting the chisel out of parallel to the timber when chopping, and not chopping vertically causing an undercut mortice.

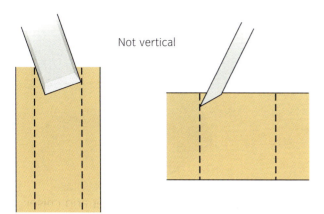

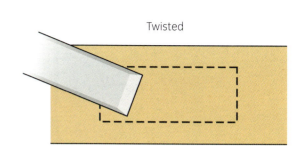

Creating wide and undercut mortices

PARING

Paring is the removal of small amounts of timber without the use of a mallet. Paring can be carried out with the grain or down the end grain. As with all chiselling operations, the timber must be **immobilised** to ensure you are not chasing the timber across the bench and to ensure safe working. The timber should be in the vice or cramped down to a suitable work surface for safety. The weight of your shoulder can be added if needed.

INDUSTRY TIP

Use a scrap of clean timber between the cramp face and the work to prevent damage to the face.

Immobilised

Held securely

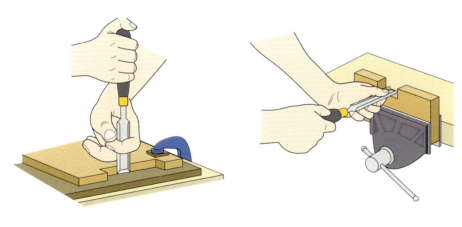

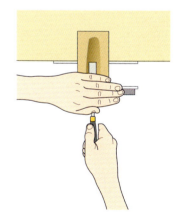

How to correctly pare with a chisel

STORAGE AND PROTECTION OF CHISELS

When chisels are purchased in sets, they will come in packaging depending on the quality and cost. This could be either a robust storage box or plastic wallet. All chisels will have a chisel guard protecting the cutting edge from damage and for safety. In use, the storage box is often used by joiners but not site carpenters as the boxes are bulky and take up valuable room. Site carpenters often use 'chisel rolls' which hold more chisels and can be laid out where they are working for quick access.

Storing chisels in a chisel roll

Chisel guard

MAINTAINING PLANE IRONS AND CHISELS

To cut safely and efficiently, planes irons and chisels must be kept sharp. A general rule is that for every hour you use these tools you should spend five minutes sharpening them. The more **abrasive** the material being cut, the more frequently you will need to sharpen them. As the cutting edge wears, it becomes blunt. The table below shows other instances when you need to re-grind a chisel:

Abrasive

The wearing nature of the material, for example chipboard is made of wood particles bound together with a synthetic adhesive. This would wear/dull cutting edges far quicker than planing softwood

Issue	Reason
Rounded cutting edge	From not maintaining a constant angle when sharpening.
Cutting angle is too steep Correct angle	Tool rests on grinder set incorrectly.
'Nicked' cutting edge	From using chisel on abrasive materials.

The effect of a blunt cutting edge is twofold:

1 You will need to use greater force, possibly leading to an accident.

2 The finish produced is poor and could damage the product being worked.

The longer you leave a cutting edge between sharpens the longer the sharpening process will take. Each time you sharpen a plane iron or chisel you will increase the sharpening angle slightly. The most efficient angle is between 25 and 30°. Once you have reached 30° it will need regrinding back to 25°. The image below shows the standard grinding angle, increasing angles of sharpening and the standard sharpening angle.

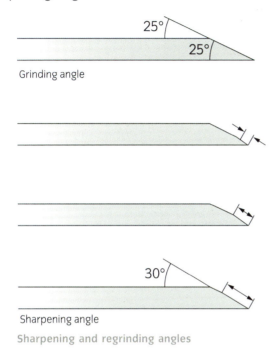

Grinding angle

Sharpening angle

Sharpening and regrinding angles

All chisels (except skew) are ground and sharpened square to the face. If, during the sharpening process, the cutting edge runs out of square (easy to do especially with narrow chisels) it will need truing again by grinding.

CHISEL GRINDING

The legislation that covers the safe use of the grinder and all power tools is PUWER (Provision and Use of Work Equipment Regulations 1997, see Chapter 1, pages 16–17, for more information). The Safety in Use of Abrasive Wheels document will also need to be conformed to. For full details of these visit www.hse.gov.uk.

Prior to using a grinder you must have been trained in its use. There will be an authorised person within your company or training centre who will have been on an abrasive wheels course, who is able to change a grinding wheel and check that they are safe to use.

ACTIVITY

Research two different types of eye protection.

PPE

The personal protective equipment required when carrying out hand tool maintenance would be:

PPE	When to use
Safety goggles	When grinding
Respiratory protection	When dry grinding
Safety boots	All occasions
Apron or warehouse coat	When grinding
Barrier cream	When grinding

Methods for grinding chisels

Chisels can be ground on two types of grinding machine:

1 Traditional bench grinder

The traditional bench grinder would be the most common method of grinding chisels. While they grind the metal away quickly, as it is a dry stone, the temperature will build up. If not careful, the thin tip will turn blue, losing its **temper** and softening the steel, preventing it from holding a keen (sharp) cutting edge. To prevent this, the chisel is regularly dipped in water to cool the blade.

Temper

Tempering is a process of heat treating, which is used to increase the toughness of steel

Bench grinder

The following method would be used:

1 Fill a container with cold water.

2 Set the tool rest to grind at 25° and to within 3mm of the abrasive wheel.

3 Set the transparent grinding shield and put on eye and respiratory protection.

4 Switch on the grinder.

5 Dip the chisel into water.

6 Holding the chisel against the tool rest, bring the cutting edge gently into contact with the grinding wheel.

7 **Traverse** the chisel across the full width of the grinding wheel five times, then dip the chisel in the water.

Traverse

Move from side to side

8 Repeat as above until the complete face has been re-ground to the point of the chisel.

2 Water cooled sharpening system

These are proving very popular now as an alternative to traditional bench grinders and offer a number of advantages:

- It is safer to use as it does not run at high speed.
- The stone runs through a water-filled reservoir so the chisel will not heat up and lose its temper.
- It has a sliding tool holder to improve the accuracy of grinding and safety in use.
- It is very easy to adjust the grinding angle to 25°.

Water cooled sharpening system

PLANE IRON GRINDING

This is virtually the same process as above with the same grinding and sharpening angle. The only difference is that some plane irons have a shaped cutting edge to suit different purposes.

Below are three common cutting edges and the planes they are used on. Although these are the traditional shapes, the slightly convex cutting edge proves to be the most popular.

ACTIVITY

Select two items of PPE required when grinding and state why they are necessary and what they protect against.

Type of cutting edge	Planes used on
Square	Try, shoulder, rebate, bullnose, plough and router

Type of cutting edge	Planes used on
Slightly convex ⟂ 1mm	Block, jack and spokeshave
Softened corners ⟂ 15mm	Smoothing plane

SHARPENING PLANE IRONS AND CHISELS

This can be carried out on the following range of sharpening stones

Oil stone

Oil stone is a commonly used stone (actually manufactured stone made from cast carborundum grit). Generally oil stones are double sided and are 200mm long by 50mm wide. One side has a medium grade and the other a fine grade surface. Mostly the fine surface is used. The stone is protected and stored in a wooden box (normally made by the owner). Site carpenters' boxes are generally light and simple as they are used on site. Joiners may make a highly decorated box which would sit permanently on their workbench. The stone requires lubricating in use with light machine oil from an oil can.

Water stone

The water stone will look virtually identical to the oil stone, but the water stone is made from a natural stone. These stones remove less metal than man-made stones and polish while sharpening, providing a very keen (sharp) cutting edge. They can be lubricated with either water or oil. If used with oil you cannot revert back to water so it is best to keep to water.

The disadvantage with both oil and water stones is that they wear hollow in both directions and need regular maintenance to grind them back to a flat surface.

Diamond stone

By far the most popular choice today. While they are more expensive initially, diamond stones require little maintenance and the sharpening surface stays flat. They remove the worn cutting edge rapidly and are available in various grit grades. They are available in slightly larger sizes up to 150mm long by 75mm wide, giving a larger sharpening area. It is best to apply the water to the stone using a **misting spray.** This will control the amount of water applied.

Oil stone

INDUSTRY TIP

Wipe all residue away from the stone after each use with a clean cloth. This will prevent the surface from becoming clogged and will keep the stone working effectively.

Diamond stone

Misting spray

A bottle with a pressurising handle such as a kitchen cleaner spray bottle. Always relabel the bottle showing that it contains water

Slip stone

Slip stone

Slip stones are generally made from cast abrasive grit. This allows for a range of shapes and sizes to be manufactured. They are normally quite small, approximately 125mm long by 50mm wide.

Sharpening (honing) process

With continued use, all cutting edges become blunt (dull) and will need honing. Dull cutting edges require extra force and produce poor results so to keep cutting edges keen (sharp) follow the following process:

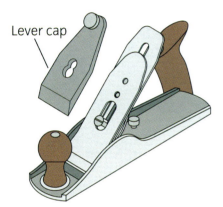

Lever cap

STEP 1 Remove the lever cap.

STEP 2 Remove the plane iron assembly.

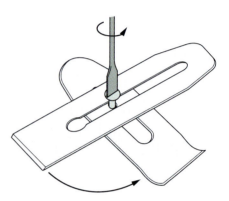

STEP 3 Remove the back/cap iron safely.

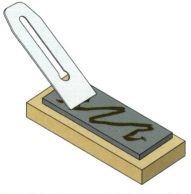

STEP 4 Apply appropriate lubricant to the stone. Spread the lubricant over the stone with the wire/burr edge chisel or plane iron (blade).

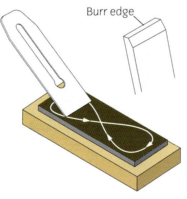

Burr edge

STEP 5 Place the sharpening edge of the blade on the stone and find the sharpening angle. Push the blade up and down the stone maintaining this angle whilst covering the whole stone. Stop when the worn cutting edge has been removed and burr is found on the back edge of the blade.

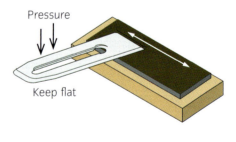

Pressure

Keep flat

STEP 6 The back edge blade is then placed flat down on the stone and rubbed up and down the stone until the burr has been lost.

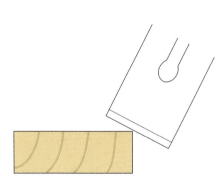

STEP 7 The final wire-edged burr can then be lost either on a leather stropping block or on the corner of a piece of clean softwood.

STEP 8 Finally wipe all excess lubricant away from the cutting edge with a clean rag or paper wipe, and reassemble plane.

SHARPENING GOUGES

Scribing gouges

Scribing gouges are sharpened on the inside face. The slip stone can either be secured in a bench vice and the gouge moved over on its top edge, or the gouge can be held down to the edge of the bench and the slip stone moved over the cutting edge (try both methods and see which you prefer).

INDUSTRY TIP

The sharpening angle is found by gently lifting the back of the blade until a seam of the lubricant is squeezed out at the tip.

ACTIVITY

Under supervision, carry out the sharpening process on a 25mm chisel.

Sharpening the inside edge of a scribing gouge

Removing the burr by rolling the back edge on a flat stone

Firmer gouges

Firmer gouges are sharpened on their outside face on a normal sharpening stone. The slip stone is used on the inside edge to remove the burr.

INDUSTRY TIP

Every carpenter and joiner at some time has used blunt planes and chisels and regretted it. It pays to sharpen tools at the beginning of every new activity, that way you will always achieve quality results.

Wheel brace

Swing brace

Cordless drill

HAND-HELD DRILLS

Pure hand-held drills are rarely used these days as their use has been replaced with cordless drills. It is important, however, to know what drills are available to use.

WHEEL BRACE

Wheel braces are available with a three pin chuck which will take a range of twist bits from 1mm up to 8mm in diameter. They are used to bore and countersink holes (see page 145), usually for screws.

SWING BRACE

The size of a swing brace is determined by the sweep of the brace (200–300mm diameter), the most common size being 250mm. They are used for boring larger holes as the turning capacity of a swing brace is more effective than a wheel brace. Most now have a ratchet facility which allows holes to be bored close to an obstruction.

CORDLESS DRILL

This is now the most common tool used to bore all holes due to its efficiency and ease of use. A variety of models are available according to price. A combination drill will be found of most use with the facility to drive screws, and bore holes in both timber and masonry (brick and blockwork). The use of this drill will be covered in more detail in Chapter 4 (page 159).

TYPES OF BIT

With the increased use of cordless drills, many older type of bits
have been adapted to be used by powered means. The examples
shown in the following table are restricted to bits in common use.

Bit	Size	Uses	Drill used in
Twist	1–13mm	Boring holes in wood metal and plastics products.	Wheel brace and cordless drill.
Lip and spur	1–13mm	Modern replacement of twist bit for cleaner cuts and accurate centre positioning (for use on timber products only).	Wheel brace and cordless drill.
Countersink	13–20mm	Producing a countersunk hole to receive the head of a screw.	Wheel brace and cordless drill.
Masonry	4–13mm	Boring holes in brickwork, blockwork and concrete for plastic plugs.	Cordless drill with a hammer action.
Screwdriver	1–3 point	Driving pozidriv and other types of screw heads.	Cordless drill on slow speed.
Jennings pattern auger	6–32mm	Boring deep holes in timber, eg boring holes for the body of locks.	Swing brace and cordless drill.
Irwin pattern auger	6–32mm	Boring deep holes in timber.	Swing brace and cordless drill.

Bit	Size	Uses	Drill used in
Centre	13–38mm	Boring shallow holes up to 50mm deep.	Swing brace.
Forstner	9–50mm	The short centre pin allows shallow (blind) holes to be bored within 3mm of the back face without showing. Suitable when fitting concealed hinges.	Swing brace and cordless drills.
Expanding auger	25–50mm	Has an adjustable wing allowing holes of any size to be bored.	Swing brace.
Flat/spade	6–50mm	Boring through softwood where a good finish is not required.	Cordless drill (although a corded drill produces better results).
Drill and counterbore	Purchased to match plug cutter.	This is a combination drill which bores a clearance hole for the screw and a hole for a wooden plug to be inserted following the screw.	Cordless drill.
Plug cutter	To produce 10,13 and 16mm wooden plugs	Produces wooden plugs which are inserted into a counterbored hole to conceal fixings.	Cordless drill.
Hole saw	16-152mm	For boring very large holes through generally thin materials (useful when running services through pre-fixed kitchen units). Sometimes known as tank cutters.	Corded/cordless drills.

BORING HOLES

Using auger bits

Auger bits were designed for the swing brace and therefore originally had a square tapered shank to fit into its alligator jaws. Their modern replacement has a hexagonal shank to fit into cordless drill chucks. They are very commonly used when boring holes for fitting cylinder and mortice locks.

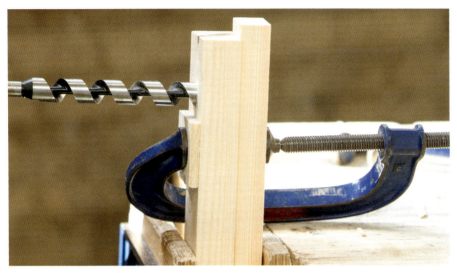

Boring a hole through timber

Using twist drills

The most common use of twist drills is preparing timber when securing two components together. The use of a cordless drill, screwdriver bit and modern screws minimises the amount of preparation required. Pilot and countersinking are now only required for hardwood or when the screw is large (10 gauge or above).

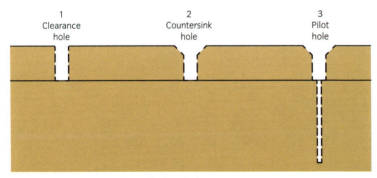

Clearance, countersink and pilot holes

It is important that the correct size of drill is selected to ensure one piece is pulled tightly to the other. A rule of thumb for drill sizes is to halve the screw gauge in mm for the clearance hole and halve that again for the pilot hole.

> **INDUSTRY TIP**
>
> A screw point pulled through candle wax or tallow (solid animal fat that can be purchased in tubs) will lubricate the screw and reduce the effects of friction when being inserted. It also minimises the corrosion and allows screws to be released more easily if required to be removed later eg on ironmongery.

> **INDUSTRY TIP**
>
> Ask a colleague to check by sight that you are drilling square to the face of the timber.

Example

Determine clearance and pilot hole sizes for a number 8 gauge screw:

Clearance hole $8 \div 2 =$ **4mm**

Pilot hole $4 \div 2 =$ **2mm**

COMMON DRILLING FAULTS

Materials can be easily damaged and work can be ruined if holes are bored incorrectly. Common faults are:

Damage	Cause
Break out (spelching) on back face	Boring through from one side only
Holes bored out of square or parallel to timber	Not aligning bit with surface
Face of timber around hole damaged	Blunt bits used
Cracked glass in door	Holes drilled too deep when fitting the mortice lock
Drill skid over surface	Drill has slipped before the hole started to bore
Hole veering off course (not straight)	Too short a drill selected for task

These are just a few common drilling faults that can cost a lot to rectify in terms of new materials or time to repair fault.

DRILLING JIGS

Drilling jigs are devices which allow repeated processes to be carried out quickly and consistently. They can be made or purchased. An example would be to consistently position door knobs and handles on kitchen cupboards.

SECURING TIMBER

To ensure safe working, timber must be secured while drilling holes. This may be in the vice, cramped to the bench or held down to a stool with your knee or a cramp.

SHARPENING DRILL BITS

Twist and lip and spur bits are normally considered a disposable item and are not resharpened.

Auger bits are sharpened using a small flat or triangular file as shown. This must be done very carefully to ensure that the spur is always proud of the cutter and the outside edge of the bit is not reduced in width when removing the burr.

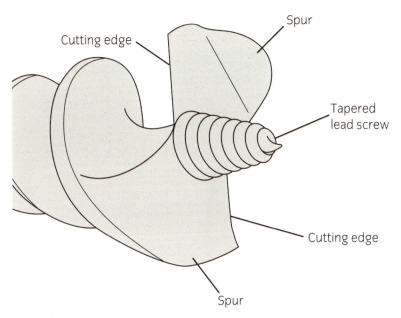

Parts of an auger bit

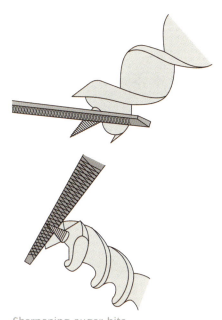

Sharpening auger bits

STORAGE AND PROTECTION OF DRILL BITS

When purchased in sets like chisels, they either come in a storage box or plastic wallet. Most carpenters and joiners use bit rolls for convenience.

Bit rolls for drill bits

GENERAL STORAGE AND TRANSPORTATION OF HAND TOOLS

Bass

A traditional tool bag that opens out flat to reveal all of a carpenter's tools

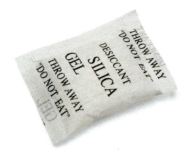

A silica gel packet

Traditionally wooden tool chests, boxes and **basses** were carried by carpenters and joiners. Today there is a range of tool bags and portable boxes that are used. They often have trays and drawers where measuring and marking tools and other small items are kept. Silica gel packets will help to stop rusting occurring during storage.

Tool storage equipment	Use/description
Tool bag	A traditional tool bag that opens out flat to reveal a carpenter's tools. Made of canvas, it is strong but light adding little to the total weight of the tools being carried. Used where only a limited range of tools are required and where access is limited. Also known as a bass.
Tool box 	Traditionally wooden tool chests were used to store tools when not in common use. Trays and compartments can be made to suit the exact size of the tools being stored. The lid was often used for the storage of large saws as shown. Tool boxes are seldom used today other than by joiners and cabinet makers who have the box stood at the end of their work bench.

Tool storage equipment	Use/description
Tool trolley 	Today there is a range of portable tool boxes available. They often have trays, drawers and compartments that can be used to separate sharp edge tools from other equipment. This will assist in quickly finding tools and prevent the sharp edge tools from damaging other equipment or becoming blunt. Tool trolleys often have lift-off boxes allowing the load to be split when the trolley cannot be wheeled.

A LAST WORD ON TOOLS

Although portable power tools are increasingly being used, the good carpenter and joiner will still need a wide range of hand tools and be very competent in their use and care. Good tools are expensive (those at the 'cheaper end' of the price range are generally inferior in quality and perfomance), but early investment will pay dividends. Carpenters and joiners will need to add to their tool kit because the nature of the work they are doing changes, new tools are available or to replace those lost, stolen or worn out. Keep an eye out for new developments as many offer advantages in terms of ease of work or time saving. Always ensure that your tools are secure at the end of the working day, they are expensive to replace.

There are many other carpentry and joinery tools not shown as they are out of scope of this chapter. These will be shown elsewhere in this and the Level 2 book.

Case Study: Matthew

Matthew was coming to the end of his first year of study at college. He had really enjoyed the course and planned on returning the following year and to actively look for work. His lecturer had always emphasised that when buying hand tools to always go for the best as they would last a long time. He had learned a lot about hand tools and found the college ones, although old, easy to use and produced good results. He was worried that he would not be able to afford tools of that quality.

After saving a sum of money, Matthew decided to buy some cheap tools from a market. He chose a selection that he thought would be sufficient to start working for friends and relatives as he built up his experience. During his first job he proudly produced these tools from a holdall and showed them to his uncle, who he had volunteered to rehang a door for.

Matthew sharpened all his chisels and plane irons and set about the task of hanging the door. He quickly found that when planing the door to fit the frame, he could only take off a few shavings before the plane became blunt and started tearing the timber up. He re-sharpened it and found the same thing happened again. His uncle spotted Matthew was having a problem and offered him an old Record plane of his own. Matthew was pleased to see this as it was the same model he had used at college. He sharpened and used this and had no further problems. He realised that buying cheap tools was a mistake and went on to buy good quality second-hand tools.

Work through the following questions to check your learning.

1 Steel tape measures are available in lengths between

a 1–3m

b 2–8m

c 5–8m

d 8–25m

2 A try square is only used to mark lines at

a 45°

b 60°

c 90°

d 135°

3 Which **one** of the following tools can be set to mark lines at any angle?

a Sliding bevel

b Try square

c Folding rule

d Combination square

4 The depth of a stub mortice is **best** checked using a

a Marking gauge

b 150mm steel rule

c 2000mm steel rule

d Sliding bevel

5 A rip saw is used to cut

a Mitres

b Down the grain

c Across the grain

d The waste from dovetails

6 How many teeth does a panel saw have per 25mm?

a 3–5

b 8–10

c 12–14

d 16–22

7 The teeth of a saw are protected when not in use by using a

a Gullet

b Saw back

c Saw guard

d Plastic bag

8 The bottom of a plane is called the

a Base

b Frog

c Sole

d Foot

9 Which **one** of the following planes is used for fine finishing?

a Try

b Jack

c Fore

d Smoothing

10 Which plane is used to plane curved surfaces?

a Block

b Plough

c Rebate

d Compass

11 Mortice chisels are available in a width of sizes between

a 3–6mm

b 6–13mm

c 10–38mm

d 25–50mm

12 Which **one** of the following chisels is used when cutting dovetail joints?

a Firmer

b Paring

c Scribing gouge

d Bevel edged

13 A mallet should be held

   a By the head

   b Under its head

   c With both hands

   d At the end of its shaft

14 Mortices should be chopped

   a From both sides

   b From one side only

   c Using a firmer chisel

   d Using a paring chisel

15 Which **one** of the following is the correct sharpening angle for plane irons and chisels?

   a 10°

   b 20°

   c 30°

   d 40°

16 A smoothing plane iron should be ground

   a Skew

   b Square

   c Slightly convex

   d With softened corners

17 A scribing gouge is sharpened with which **one** of the following stones?

   a Slip

   b Gem

   c Pumice

   d Grinding

18 The largest twist bit that will fit into a wheel brace is

   a 4mm

   b 8mm

   c 16mm

   d 20mm

19 Auger bits are sharpened with which **one** of the following files?

   a Mill

   b Round

   c Triangular

   d Half round

20 What size clearance hole would be bored for a size 10 wood screw?

   a 2.5mm

   b 5mm

   c 7.5mm

   d 10mm

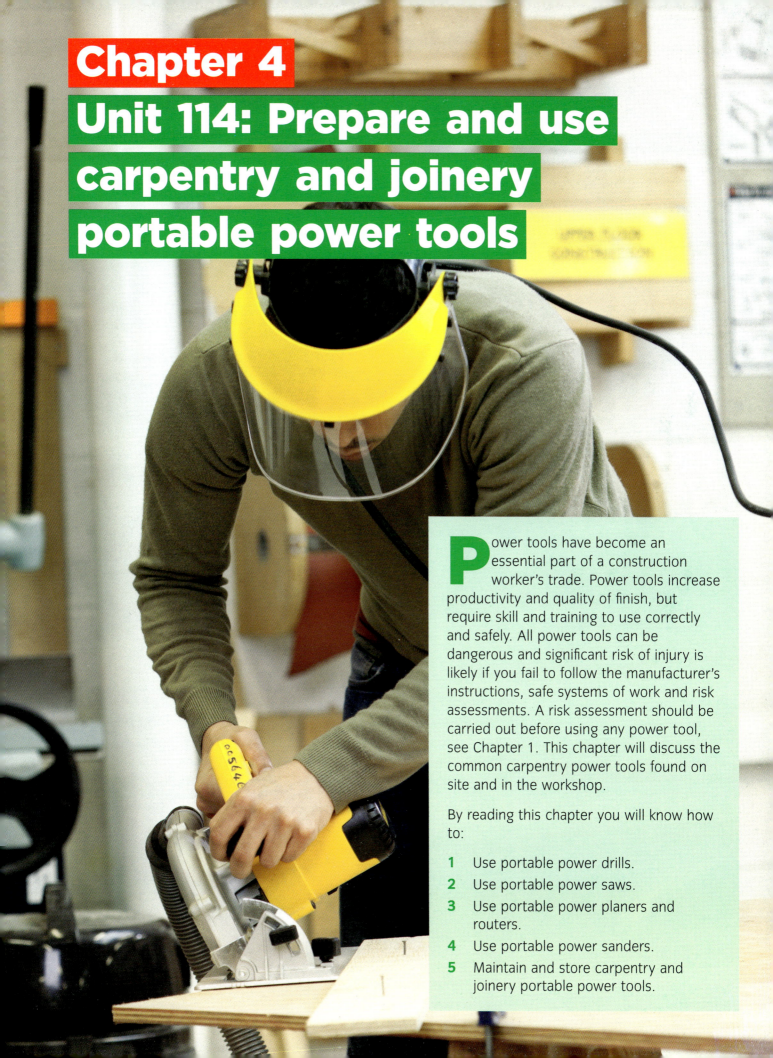

Chapter 4

Unit 114: Prepare and use carpentry and joinery portable power tools

Power tools have become an essential part of a construction worker's trade. Power tools increase productivity and quality of finish, but require skill and training to use correctly and safely. All power tools can be dangerous and significant risk of injury is likely if you fail to follow the manufacturer's instructions, safe systems of work and risk assessments. A risk assessment should be carried out before using any power tool, see Chapter 1. This chapter will discuss the common carpentry power tools found on site and in the workshop.

By reading this chapter you will know how to:

1 Use portable power drills.
2 Use portable power saws.
3 Use portable power planers and routers.
4 Use portable power sanders.
5 Maintain and store carpentry and joinery portable power tools.

POWER SOURCES

Portable power tools require a power source that can be connected via a flexible lead or can be small enough to be directly connected (for instance battery powered) to the tool to allow it to be portable. There are several types of power sources.

ELECTRICITY

Electricity is the most common form of power used for tools and equipment. It can be supplied via the mains, or stored in a battery and then connected to the tool.

Mains electricity is generated in power stations. It is delivered via power lines and, using transformers, the power (voltage) is reduced. Domestic power (as found in normal houses) is 230V (volts) and will kill a person if an accident were to occur. To reduce the likelihood of death and serious injury on site as a result of electric shock, the power supply on site is reduced to 115V. This is more commonly referred to as 110V. This is achieved using a transformer, usually yellow in colour.

Transformer

Although 110V is preferable to 230V on site, 230V tools are often used in joinery workshops. Another precaution is the use of an **RCD** which acts as a failsafe trip system. If the tool you are using develops an imbalance in the electrical current (like cutting through the cable) it then immediately cuts the electrical supply to prevent electrocution.

Hand-held power tools will have a double insulation symbol on them which means the tool is designed in such a way that the electrical parts can never come into contact with the outer part of the tool.

Within workshops 415V power supply is a common feature. The power connectors for this source are red and this type of power source only tends to be used on large fixed machines requiring lots of power, such as the large woodworking machines used in production of moulded and planed timber.

ACTIVITY

Use this web link to investigate risk assessments for power tools: www.hse.gov.uk/risk/casestudies/woodworking.htm

INDUSTRY TIP

Before changing any tooling in any power tool always make sure it is disconnected from the power supply. This will make sure there is no risk of accidental start-up.

INDUSTRY TIP

Different electricity voltage is shown using colours. Red is 415 volts, blue is 230 volts (commonly called 240V) and yellow is 115 volts. Refer to Chapter 1, page 35 for more information on voltages.

RCD

Residual Current Device

INDUSTRY TIP

If using a power tool for the first time, make sure that you have been trained first, and are supervised as appropriate.

Double insulation symbol

A transformer needs to be as close to the 230V supply as possible, therefore minimising the exposure to a higher voltage.

Leads need to be kept neat and tidy. Care needs to be taken to make sure leads do not become a trip hazard, and are also protected from damage. If possible try to keep any running lead above head height to reduce the chance of accidents or damage to the lead.

Any extension lead needs to be fully uncoiled to make sure the cable does not overheat and become a fire hazard.

Lead protection

Make sure the leads are not a trip hazard

INDUSTRY TIP

Remember to keep the transformer as close to the 230V supply as possible and use an 110V extension lead if required.

BATTERY

Battery powered tools do not require a connection to a mains supply, and this does have advantages with regards to safety and convenience. However, they tend to be more costly, especially the batteries which are usually the most expensive part of the tool. Also, the batteries do require charging, and as the chargers are 230V they need to be kept somewhere safe and dry.

An 18V and a 12V battery

Batteries are available in a wide variety of types and voltages, the most common ones are 12V, 18V, 24V and 32V, but there are many more. The higher the voltage, the more energy the tool will have. Batteries are also marked with Ah (ampere hours); the higher this figure the longer the battery will last.

A power drill with a battery pack attached

Gas canister

GAS

Nail fixing tools (commonly known as 'guns') are often powered by gas and battery. The gas is stored in a small canister that is loaded into the tool. A combustion chamber is filled with the gas and then ignited using a spark provided by the battery. Gas canisters (even empty ones) require careful handling and must be disposed of correctly (not on a bonfire!).

COMPRESSED AIR

Compressed air is used to power a variety of hand tools. High pressure air is produced in a compressor, where it is stored in a tank. Pipes and hoses take the air to the tools. The advantage with compressed air is that there is no electricity involved other than the power needed to run the compressor. However, air powered tools can be very noisy and the leads can get in the way as they are thick and rather inflexible compared with electrical leads. Compressed air itself can be extremely dangerous: if misused the air can enter the bloodstream.

Compressor

Lead

Air drill

Air sander

POWER TOOL SAFETY

The legislation that cover power tools (HASAWA, PUWER, COSHH and the Control of Noise at Work Regulations) is covered in detail in Chapter 1. It is important that these are followed as power tools can cause serious injury if not maintained or used without proper training.

Appropriate PPE must be worn when using power tools due to high noise levels, vibration, dust and shavings produced and other potential flying debris. Full details of these items are given in Chapter 1.

ACTIVITY

Take a power tool and inspect it. What safety features does it have? Is it in good condition?

LOOKING AFTER POWER TOOLS

Power tools are expensive, and careful handling and storage is required if they are to have a long life and remain safe to use. Tools are often supplied in purpose-made boxes or cases, which have spaces for attachments, accessories, fences and guards. Using these boxes to store tools after use will help prevent damage and loss of parts. Tools should be checked before and after use for damage. A manufacturer's booklet is provided when power tools are supplied. This will carry safe operation instructions as well as maintenance details.

WHAT TO CHECK FOR

Power tools can suffer from the following faults:

- missing or damaged guards and fences
- damaged casing or handles
- faulty wiring and plugs
- faulty switches
- blunt or otherwise dangerous tooling.

Tools must not be used if they are faulty or if temporary repairs have been carried out.

Damaged power tool lead

Worn router cutter

Bent jigsaw blade

Blunt teeth on sawblade

PAT (PORTABLE APPLIANCE TESTING)

PAT testing is a regular test carried out by a qualified electrician to make sure the tool is in a safe electrical condition. A sticker with the date of the test is placed upon the tool when tested, and tools that do not pass are taken out of use.

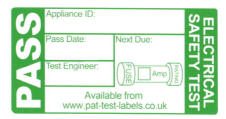

PAT testing labels: fail and pass

POWER DRILLS

Power drills are probably the most common power tool available. They range in size from large **SDS** drills designed to drill stone and concrete down to small drivers used for inserting small screws into hard-to-reach places. Drills can be powered by mains electricity, battery or air. Battery drills are usually **dual purpose** in that they are designed to drill holes and drive screws, and these are commonly referred to as drill-drivers. The following images show the parts of a battery drill driver, then the parts of a corded SDS drill.

SDS

Slotted Drive System – a system that allows quick change of bits and efficient masonry drilling

Dual purpose

Designed for two purposes

Gear selector

Clutch selector (for not damaging screw heads)

Keyless **chuck**

Trigger

Battery

Chuck

The part of a power drill that holds the rotating part

SDS chuck

Depth gauge

Auxiliary handle (to be used when heavy drilling)

Trigger

Repetitive

Many of the same

Another commonly found drill within workshops is the pillar drill. In the simplest terms it is a drill incorporated into a stand holding the drill head, a table that usually is able to rise and fall which the components to be drilled can be attached to, and a handle which will lower the drill head. These are particularly useful where **repetitive** holes are required to be drilled accurately

Pillar drill

SCREWDRIVER BITS

There is a wide variety of screw head designs, and it is important to choose the right screwdriver bit for the job as using the wrong bit type or size will damage the screw head. Below are three types of screwdriver bits.

Screwdriver bit and head view: Pozidriv (P2), Phillips (PH) and slotted

Commonly used screwdriver bits come in three sizes: 1, 2 or 3, 3 being the largest.

ACTIVITY

Look closely at a pozidriv and a Philips screw. Sketch the screw heads. What is the difference between the two?

The following are the four types of chuck: keyed (with key), keyless and SDS, and an impact driver with hex fitting.

Keyless

Keyed

SDS chuck

Impact driver with hex fitting

DRILLING

When drilling through material, it is important to make sure that there will be no break out. This can be prevented by boring from both sides or by using a block of timber fixed to the back side by a cramp if possible.

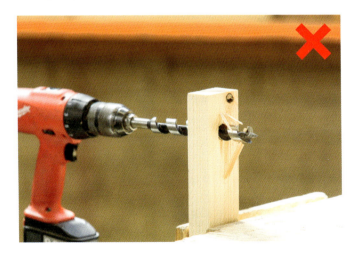

DRILL BITS

There are several types of drill bit designed for different tasks and materials. For information on drill bits such as their size, use and photographs, see Chapter 3, pages 143–144.

FIXINGS

Before fixing to a wall or into a floor, it is important to locate any hidden services such as water or gas pipes or electricity cables so they do not get damaged. There are several methods of locating services. A detector will locate pipes and wires. Wires should run vertically up or down from a socket, so it is very likely there will be a hidden wire buried in the wall in these areas.

Detector

Wires running vertically from socket

The most common two types of wall that are fixed into are solid (such as concrete or brick) and hollow (such as stud partition). Different fixing methods are required for each type.

Fixing	Image	Wall type
Rawlbolts		Masonry
Coach screw		Timber
Raised screw		Metal
Countersunk screw		Timber

Wall anchor for hollow walls

Fixing	Image	Wall type
Round screw		Timber
Bugle headed screw		Plasterboard into timber
Cavity fixing		Plasterboard
Plug and screw fixing		Masonry
Chemical fixing		Masonry

ACTIVITY

You have been asked for guidance on selecting the correct equipment needed to fix a timber batten that is 1200mm x 50mm x 25mm, to be fixed into a brick wall to hold up a curtain rail. List the tools required for this task.

CLEARANCE, COUNTERSINK AND PILOT HOLES

A clearance hole helps the joint pull tight as the screw thread isn't holding in the piece being fixed and the screw head will pull it tight. Countersinking will give the screw head a neat finish upon completion and also prevents splitting. A pilot hole prevents wood splitting.

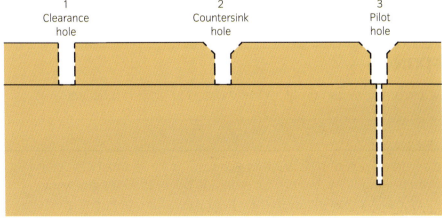

Clearance, countersink and pilot holes

POWERED NAILERS

Cordless nailer

Angled gas brad nailer

There is quite a variety of power nailers using various power sources including electricity, gas and compressed air. It is very important to read the manufacturer's instructions and to be trained before using a nailer (as it is with any power tool). They come in various sizes, from small pinners designed to fire small fixings such as pins or staples into thin plywood or beads, up to large nailers which can drive large (up to 100mm) nails into **carcassing** or even into steel. The parts of a nailer are labelled below.

Carcassing

Studwork or roofing/flooring timbers

> **INDUSTRY TIP**
>
> Ensure the nailer cannot fire by removing the power source before any type of maintenance is carried out on the tool, including removing jammed nails. Reloading is usually a safe operation to be carried out while connected to the power source. Always refer to the manufacturer's instructions first.

Safety device (will only fire if depressed)

Trigger

Nail magazine

Spring loaded lever

Remember: always wear suitable eye and ear protection at all times when using this type of power tool, even if you are only going to insert one fixing.

FIXINGS

Below are common types of fixings used in powered nailers.

Fixing	Use
Nails	There are several types of nails used in guns, some have small heads and are used for finishing work, others have bumps down their length, known as 'ring shank nails' which increase their hold. Nails for use in nail guns are usually collated - this means they are stuck together either with paper or glue. This makes it easier and quicker to load the nails into the gun.

Fixing	Use
Pins or brads 	As with nails, these are usually collated. Used for smaller joinery jobs.
Star dowels	Used to fix mortice and tenon joints together. After cramping the joint, these are fired through the face of the joinery into the joint. These are made from light alloy, meaning they will not damage tools when cleaning up.
Staples	Used for joinery, often for fixing thin plywood.

CRAMPING WORKPIECES

There are several devices available that can be used to secure pieces while you are working. The table below shows four different cramps you could use.

F Cramp
Often referred to as the speed clamp. This cramp has an adjustable arm which can be set to accommodate any width with the distance of the main bar.

G Cramp
Available in several sizes and depths and they can be very useful in applying enormous amounts of pressure.

Quick release cramp
For light, temporary jobs, such as holding down work. They are not very strong, so are not suitable when gluing.

Sash cramps
Like quick release cramps. When placed around timber and then put in a vice, you can machine along the edge of the timber without catching on the vice.

POWER SAWS

There is a wide variety of power saws available. This book will deal with three types: chop saw, hand-held circular saw and jigsaw.

CHOP SAW

Also known as a mitre saw, this is a very useful tool used for making accurate square, angled and **compound cuts**. The following image shows the parts of a chop saw.

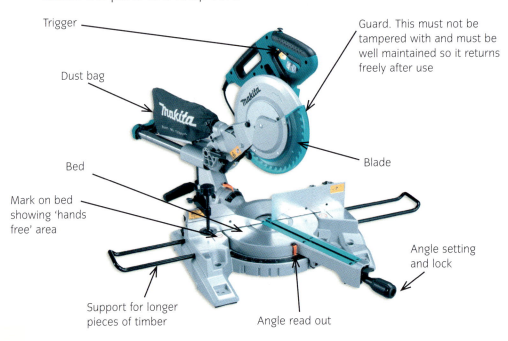

Trigger

Guard. This must not be tampered with and must be well maintained so it returns freely after use

Dust bag

Bed

Blade

Mark on bed showing 'hands free' area

Angle setting and lock

Support for longer pieces of timber

Angle read out

Architrave mitre

There are angle adjustments on the bed and at the back that allow the saw to **cant** or swivel around. Some chop saws have a slider which allows wider boards to be cut. Chop saws are on site for a wide variety of tasks including cuts to **architraves** and skirting, stair spindles, studwork, rafter cuts and any other job where a straight clean cut is required.

Chop saws are now a common feature around most workshops and building sites. There is a tendency for all workers to think they are easy and safe to use and they don't need any training on them, but like all power tools they are dangerous and safe working practices must be followed to avoid serious injury. Extreme care should be used at all times while using these types of machines. When the correct training is received and followed the likelihood of injury is minimal.

As with all power tools, before you use them check the machine. Is it safe to use in its present position? Is it in good working condition? Have you been trained and authorised to operate it? Do you have the correct PPE?

While operating the chop saw keep your hands well clear of the saw blade: this means not putting your hands in the 300mm semi-circular zone in front of the saw blade while cutting. Remove short ends and off-cuts with a **push stick**. Holding short pieces is done either with a clamp, often supplied with the machine, or another suitable clamping device.

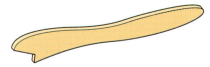

Push stick

A piece of timber at least 450mm long

Push stick

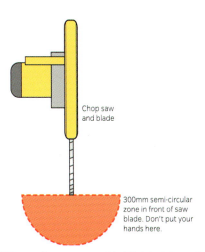

Chop saw and blade

300mm semi-circular zone in front of saw blade. Don't put your hands here.

A chop saw with the 300mm danger zone highlighted in red

If you need to cut long lengths of timber, an end support must be fitted. This prevents the timber from tipping over and so prevents the risk of injury or damage. Safe examples of these can be supplied from the manufacturer of the chop saw.

A stand used with a chop saw for long lengths of timber

OUR HOUSE

You have been asked to fit the spindles to the staircase in 'Our House' and will be using a chop saw. Describe how you would set up the job so it could be carried out safely. How many spindles will be required?

TCT circular saw blade

Ripping

Running with the grain

Metals used for tooling

Tooling can be made from several types of metal:

- *TCT (tungsten carbide tipped)*. Tungsten carbide is very hard and keeps its edge (stays sharp) very well. However, it is very brittle so it is usually bonded onto a steel body as a hardwearing tip.

- *HSS (high speed steel)*. A cheaper material than tungsten carbide, but does not keep its edge as well.

- *PCD bits (polycrystalline diamond)*. These have industrial diamonds bonded onto the surface which make them very hard wearing.

CHANGING CIRCULAR SAW BLADES

This should only be carried out by a trained competent person. Rip saw blades with a positive hook are used for **ripping** and must not be used with a chop saw. Crosscut blades are used in a chop saw. Smaller teeth give a cleaner cut. Combination blades are suitable for ripping and crosscutting. The correct blade types for crosscutting are neutral and negative, and positive for ripping only, see below:

From top to bottom: positive, neutral and negative blade types

Changing a saw blade

STEP 1 After removing the plug from the power supply, undo the nut that holds the blade. Make sure you use the spanner supplied with the tool.

STEP 2 Carefully remove the blade and store it safely for resharpening if required.

STEP 3 Carefully open the package (if you are using a new blade).

STEP 4 Read the manufacturer's instructions, checking that the blade is suitable for use with the tool and what the tool will be used for.

STEP 5 Carefully replace the blade, making sure it is seated properly into the machine.

STEP 6 Replace the nut.

STEP 7 Tighten with an appropriate spanner.

Use of a circular saw

Use the fence on a circular saw for safety

Use a straight edge for a straight cut

INDUSTRY TIP

Do not interfere with the guarding – it's there for your safety! Any damage to the tool must be reported, and repaired before the tool is used again.

Bind

When a saw blade jams

HAND-HELD CIRCULAR SAW

A versatile tool used to cut sheet materials. The correct blade must be selected (see saw blades on page 168). These saws are liable to **bind** (get stuck in the timber being cut if it is damp, not supported properly or twisted) if not used with care. The following image shows the parts of a hand-held circular saw.

Guard

Depth adjustment. The saw should be set so the blade just clears the underside of the material being cut

Retractable guard

Motor

Angle adjustment

Blade

Bed

Most models of hand-held circular saw have a riving knife. This is a piece of metal behind the blade following its curve and should be thicker than the blade, but smaller than the **kerf**. Its purpose is to stop the timber which has just been cut from closing on the back of the blade and causing the saw blade to bind. If this happens it could cause the saw to jump in the cut or even burn out the motor. In use, the saw depth should be set so that the teeth of the saw just project through the timber to make cutting safe.

Kerf

Width of the cut

Blade/riving knife

Blade teeth just projecting through timber

Fence in use

Aperture in a worktop of an unfinished kitchen

JIGSAW

A versatile power tool used to cut shaped work (curved lines). The type of work undertaken might include cutting out hob and sink **apertures** in worktops or curved shapes from plywood. The image on the following page shows the different parts of a jigsaw.

Aperture

An opening, eg in a worktop for a hob to be fitted

Trigger

Motor

Pendulum setting. The higher the, setting the faster the cut

Blade release

Bed

Blade

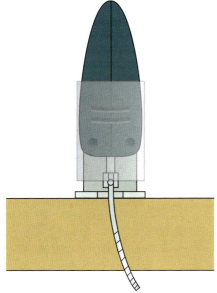

Undercutting blade

Jigsaw blades that have become dull can tend to bend inside the cut, so it is important to check this while cutting, especially when cutting thicker material such as kitchen worktops. A good way of ensuring a straight cut is to make sure the blade is regularly changed.

Blade clamp on jigsaw

Many jigsaws have a pendulum action. If this is set high the cut will be fast but less accurate. This setting allows for a faster cut as the blade clears on the down stroke, but due to the pendulum action does not allow for the cutting of tight (small radius) curves. Additionally a fast setting will increase the chance of the blade undercutting. If it is set low it will cut less quickly as the blade will rub against the timber on the down stroke but allows for the cutting of tight curves and gives a finer cut.

PROFILES

There are two types of blade profile: up cutting and down cutting profiles.

Up cutting profile

These types of blade cut on the up stroke of the jigsaw and they have a tendency to chip the upper surface while cutting. Use this type of blade when the underside needs to have a good quality finish.

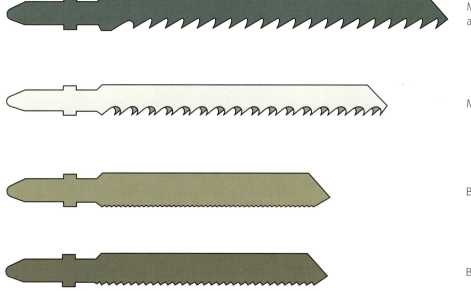

Multi purpose blade used to cut timber and board materials

Multi purpose blade for timber and plastics

Blade for cutting aluminium

Blade for cutting thin steel

Down cutting profile

Down cutting blade profiles cut on the down stroke of the jigsaw and are used when the upper surface needs a good clean cut. A good example of when to use this type of blade profile is a **laminated** worktop that is being cut with the laminate facing upwards.

Laminated

A plastic sheet glued on top of a base layer like chip board, as used in kitchen worktops

Down cutting blade

Below are some examples of the different uses for a jigsaw.

Free hand cutting

Free hand following a curved line

Following a batten fixed down

Cutting out a worktop using a down cutting blade

Cutting laminated surfaces

Straight cut

PLANERS

Planers are used to plane timber to produce a flat surface, reducing it in thickness or width. Planers are available in corded and battery powered versions. The parts of a typical planer are labelled below.

Depth adjustment

Trigger

Bed

Fence (for cutting rebates)

Planers can also produce shapes such as bevels, chamfers and rebates. A splay is an angle formed which extends the full width of the face. A chamfer is an angle formed across one edge of a piece of timber generally at 45°, and a bevel is similar but the angle generally isn't 45°. A rebate is a rectangular groove or recess in the edge of a board which holds a panel or glass in a door or picture-frame.

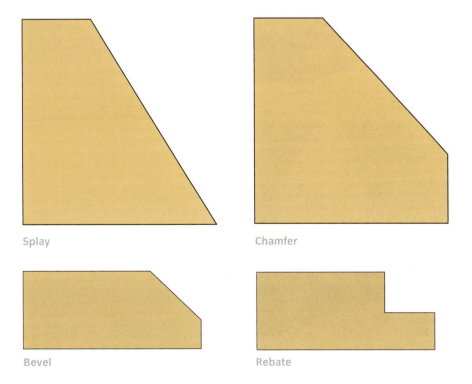

Splay

Chamfer

Bevel

Rebate

Planing the edge of timber

Producing a chamfer

Producing a rebate

The blades in a planer are known as knives. The knives commonly used on planers are disposable. The following step by step guide explains how to change a knife in a planer.

STEP 1 Loosen holding bolts.

STEP 2 Slide knives out.

STEP 3 Clean resin.

STEP 4 Replace new knives (or reverse old knives).

STEP 5 Tighten bolts.

Turning the front knob on planers lifts and lowers the front bed and so adjusting the amount of timber to be planed off in each pass of the tool. The slower the tool is passed over the workpiece, the better the finish will be. As the knives (blades) rotate, small 'scoops' are taken from the surface during planing. These marks are referred to as pitch marks. The closer they are together the better the quality of finish.

Pitch marks made by rotary cutting tooling. The left hand piece shows a slower feed speed, better finish and small pitch marks. The right hand piece shows a faster feed speed, poor quality finish and big pitch marks.

When working with planers, remember to be tidy! Keep your work area free of shavings and dust as this is hazardous – dust can cause breathing problems and be slippery. It is surprising how quickly debris can build up. One piece of equipment you should use is **LEV**, local exhaust ventilation. Commonly referred to as extraction, this equipment vacuums away dust from the work area, and is usually attached to the tool being used. Additional extraction can be supplied through a special slatted work table. Bags are supplied with most planers, however these are less efficient than LEV.

ROUTERS

Routers are very useful power tools, and have become indispensable to the wood trades. A very versatile tool, it can form **grooves**, rebates, **housings**, openings and **profiles**. Essentially, a router is a big motor attached directly to a cutter via a collet (see the bulleted list on the next page). Fences, guide bushes and stops control and adjust the depth and width of cut.

The parts of a router are labelled on the following page.

ACTIVITY

Using a power plane, plane the surface of a piece of timber on one side with a fast feed speed and on another side using a slower feed speed then compare the quality of the surface finish.

INDUSTRY TIP

If you get tram lines (raised lines down your cut) it means you have damaged cutters and they need changing.

LEV ready to use

LEV

Local Exhaust Ventilation

Grooves

A narrow cut or channel

Housings

Joint consisting of a groove usually cut across the grain, into which the end of another member is housed or fitted to form the joint

Profiles

A shape cut into timber along its length. Examples would be a rebate or ovolo

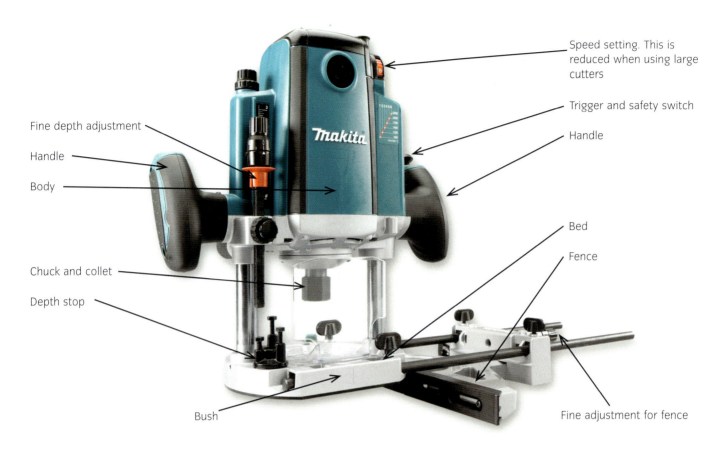

Fine depth adjustment

Handle

Body

Chuck and collet

Depth stop

Bush

Speed setting. This is reduced when using large cutters

Trigger and safety switch

Handle

Bed

Fence

Fine adjustment for fence

The following list explains the parts of a router.

- *Speed setting.* This can increase or decrease the speed to the motor. A larger cutter will require a slower speed for safety reasons.

- *Switch.* This is used to operate the machine. It is important not to start the machine if the cutter is in contact with the wood being cut. Most machines now have a slow start, meaning there is no 'kick' and all new machines are fitted with a braking system to make sure the cutter stops within a few seconds of the switch being turned off.

- *Handles.* These are used to firmly grip the tool. In some smaller models, twisting the handle acts as a depth lock.

- *Body (motor).* This is the casing that holds the motor.

- *Collet.* A chuck that holds the router cutter.

- *Bed.* The base of the tool that is in contact with the work.

- *Depth stop.* This is used to control how far the router will plunge, that is depth of cut. A castellation/turret enables several depth adjustments to be set at the same time the castellation is turned to obtain each setting. When the desired depth is set, the depth lock will lock the bed into position maintaining an accurate depth of cut.

- *Fence.* This is used to adjust the width of cut. It is adjusted by loosening and tightening the nuts. Fine adjustment can be achieved by using the knob.

Below are also some of the accessories you will need for your router:

A router with accessories

Bush

CUTTERS

The following are different types of cutters for routers; their name explains what type of cut you would use them for.

Ovolo cutter	Rebate cutter	Chamfer cutter	Bullnose cutter	V cutter
Dovetail cutter	Trimming cutter	Flute cutter	Ogee cutter	Cove cutter

Jig

A device for guiding a tool or holding a workpiece ensuring accurate cuts

The list of work that can be done with a router is extensive and with the use of proprietary (shop bought) and own made **jigs**, many more can be added to the list (see page 183). These will include: worktop jointing, door hinge and lock cutting, letterbox cutting, openings in worktops (apertures) and circle cutting.

Operating the router usually takes one of three forms:

1 Working along the edge of the product either with a fence or the bearing of the cutter running along the product. This is the most common type of use.

Two examples of working along the edge of a product

2 Following a template. This is a guide/collar that is fixed to the bottom of the router that follows the shape of the template.

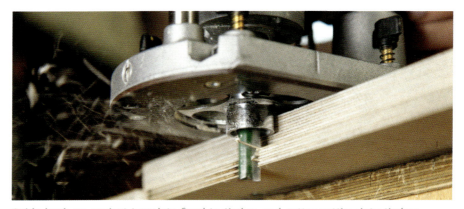

Guide bush up against template fixed to timber and cutter cutting into timber

3 A router table. This is where the router is held inverted (upside down) in a table with fences that your timber can run against.

A router table

The types of material that router cutters need to cut are varied but fall into two main categories:

- **Hard and abrasive materials** – such as oak, ash, beech, iroko and mahogany. Also includes man-made materials like plywood, MDF (medium-density fibreboard), chipboard and laminated materials.

- Softwood – such as European redwood, spruce and cedar. For more information on softwood, see Chapter 5 pages 192–193.

Like chop saws, router cutters come in the following three types of cutting material:

- TCT (tungsten carbide tipped) (the most commonly used)

- HSS (high speed steel)

- PCD (polycrystalline diamond).

Tungsten cutters are the most suitable for hand-held routers; they are reasonably cheap, readily available and very good for cutting hard abrasive materials as well as softer materials.

HSS cutters are suitable for softwoods but must not be used for the harder abrasive types of material as they burn out the cutting edge very quickly, resulting in an extremely poor finish and a ruined cutter.

PCD cutters can be used for both soft and hardwoods but are extremely expensive and as a result are used on high volume production CNC (computer numerically controlled) routers.

Below are examples of recommended maximum depth of cut for diameter of router cutters taking a full cut in one pass:

Cutter diameter	Maximum depth of cut for each pass
6mm	3mm
10mm	5mm
12mm	6mm
16mm	8mm
20mm	10mm

INDUSTRY TIP

The use of non-slip mats is ideal for flat boards to stop the timber from moving.

If you need to work on a thin edge clamp, place a packing piece alongside to give extra support to the base of the router.

Hard and abrasive materials

A material that will quickly dull/blunt the cutting edge of tooling

INDUSTRY TIP

Although chipboard seems a bit soft, it is very abrasive (due to adhesive used in the manufacturing process) and therefore TCT should always be used when routering post-formed (chipboard) worktops. TCT is hard wearing but very brittle, be careful not to hit it with any type of metal or the cutting edge will break.

Although the chart on the previous page is a good guide, several factors can affect the maximum cutting depth you should attempt in one pass. These include:

- Type of timber: is it hardwood or softwood?
- Is it hard, abrasive timber?
- Does it have lots of knots or splits along the cutting edge?
- Feed resistance and drop in cutting speed.

If the timber has some of these factors then a smaller depth of cut will be required. Remember, if it feels hard to control, vibrates about and has excessive noise levels, you are probably trying to take too large a cut in one pass, so reduce your depth of cut. In the end you will achieve a much better finish, not damage your router and reduce the likelihood of an accident.

INDUSTRY TIP

Remember, a blunt and damaged power tool is a dangerous power tool.

INDUSTRY TIP

Watch out for loose knots as they may fly out during machining.

ACTIVITY

Practice changing a router cutter, following the step by step guide.

CHANGING ROUTER CUTTERS

STEP 1 Lock spindle shaft (or use small spanner on shaft).

STEP 2 Loosen collet.

STEP 3 Remove cutter.

STEP 4 Choose a cutter to go into the router. Make sure that the router bit shaft matches the collet.

STEP 5 Replace cutter.

STEP 6 Tighten nut on collet.

JIGS

The following image shows a guide bush (collar) in the base of a router and cutter. The jig needs to be constructed larger to allow for the difference between the two diameters.

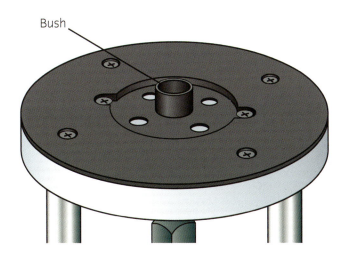

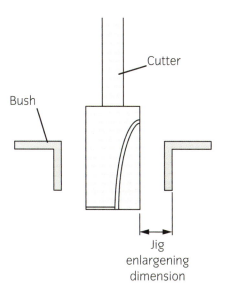

Below are examples of four types of jig you can use with your router.

Worktop jig	Hinge jig
Trammel jig in use on a router	Dovetail jig

OUR HOUSE

Take a look at the fascias, barge boards and soffit on 'Our House'. Which power tools would you need to install there? What would be the safest means of access while you carry out this work? How much material will be required?

SANDERS

There are two main types of portable sanders, those for heavy duty stock removal (cleaning up large jobs) and those for surface preparation before a finish is applied such as paint or varnish.

BELT SANDER

Welt/dwell mark

A dip in the surface of a finished product caused by a sander

Belt sanders are larger machines that feature an abrasive belt. These are very efficient at removing unwanted material, eg where there is a big step in a joint too big to reduce with a plane. Belt sanders need to be used with care because if they are not kept moving over the surface a **welt** (dwell mark) can be created, damaging the workpiece. Additionally, belt sanders are well known for catching loose clothing – use with care!

These sanders can be used for finishing after planing if a finer belt is used. More commonly an orbital sander would be a better choice because it is easier to control the amount of material being removed. Both types have their drawbacks, explained in more detail later. The parts of a belt sander are labelled below.

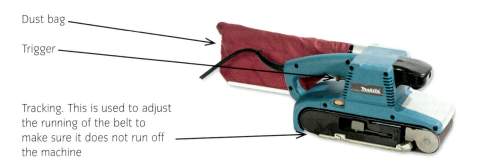

Dust bag

Trigger

Tracking. This is used to adjust the running of the belt to make sure it does not run off the machine

CHANGING A BELT

STEP 1 Loosen lever.

STEP 2 Remove belt.

STEP 3 Replace belt. Note direction of the belt from the arrow.

STEP 4 Tighten.

STEP 5 Use tracking adjustment.

ABRASIVE PAPER GRADES

Abrasive paper is graded from very fine to coarse by number of grits or particles per square 25mm. Therefore P60 (60 grits per 25mm square) is coarser than P240. Coarse grade is for rougher work, finer grades for finishing. The table below shows the different paper grades and their uses.

Paper grade	Uses
P60-P80	Rougher work, stock removal.
P100-P120	General fine finishing work.
P180-P240	Very fine finishing, generally with hardwoods.

ORBITAL SANDER

An orbital sander is a smaller machine than the belt sander and is used for finishing after the surface of the timber has been planed. You can use an orbital sander or a random orbital sander.

INDUSTRY TIP
Remember, all sanders produce fine dust particles so always use appropriate dust collection measures, even if only using for a short time. LEV is more effective than using a bag.

Front knob

Trigger

Dust bag

Most sanders have hook and eye fastenings so the pads just peel off. Some older sanders require the paper to be cut and held in clamps.

Hook and eye fastenings on sander

Paper held in clamps on a finishing palm sander

ADVANTAGES AND DISADVANTAGES OF TYPES OF SANDERS

Type of sander	Advantages	Disadvantages
Belt sander 	■ Removes large amounts of material quickly. ■ Heavy, requires little downward pressure. ■ Papers tend to last longer. ■ Good for large areas.	■ Heavier. ■ Harder to control. ■ Can leave 'welts' or dips in the surface if not kept moving. ■ Difficult to produce a good flat finish (unless a framed base is fitted).
Orbital sander	■ Easy to control. ■ Easier to control amount of material to be removed. ■ Smaller and more lightweight. ■ Tends to provide a better quality finish.	■ Difficult to remove large amounts of material due to its size. ■ Can leave small circles on the surface finish (unless a random orbital sander is used).

Type of sander	Advantages	Disadvantages
Random orbital sander	■ Easy to control. ■ Very good quality finish, virtually scratch free. ■ More flexible pad allowing better contour sanding. ■ Easy changing of sanding paper. ■ Due to light weight, easy to use in awkward positions.	■ Difficult to remove large amounts of material due to its size and weight.

All power tools are extremely dangerous if they are not used correctly, maintained according to the manufacturer's instructions and the tooling is kept sharp.

INDUSTRY TIP

Always connect your power tool to a portable dust extractor if at all possible. Dust can cause serious breathing problems if you are exposed to it over a period of time.

STORING POWER TOOLS

If power tools are stored so that cables, guards, tooling, plugs and other parts of the body are not protected then damage to these tools will be inevitable.

A little time spent storing your power tools correctly when finished with them will save time in the long run and a great deal of expense, both in replacing the power tool earlier than otherwise necessary and constantly replacing tooling that has become damaged.

ACTIVITY

You have been tasked with cutting a hole in a laminated chipboard worktop. Before you start the task, list the tools you would use, the process and all safety equipment required.

A power drill stored in its box safely

Case Study: Karl

Karl has been asked to make the section of cill below:

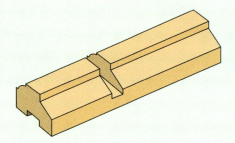

It features a housing joint, a sloping cill, an ovolo moulding and a mortar key. The cill is to be made using power tools using only ready-prepared PAR (planed all round).

- What tools should Karl use to produce the section?
- What additional equipment would be required?
- In what order would the operations be carried out?

Work through the following questions to check your learning.

1 What does the C in TCT stand for?

   a Carbon

   b Carbide

   c Carbite

   d Carbitron

2 What does the term P60 refer to on the back of an abrasive sheet?

   a Blade sharpness

   b Measure of grit

   c Type of drill bit

   d Planer blades

3 On which **one** of the following power tools would a collet be found?

   a Planer

   b Sander

   c Jigsaw

   d Router

4 What voltage is recommended to be used on site?

   a 100V

   b 110V

   c 220V

   d 240V

5 On which **one** of the following power tools would a riving knife be found?

   a Circular saw

   b Chop saw

   c Planer

   d Router

6 Which **one** of the following is the correct tooth hook for a blade used when ripping on a hand-held circular saw?

   a Zero

   b Neutral

   c Positive

   d Negative

7 Which **one** of the following power tools would be used to produce a compound cut?

   a Drill

   b Router

   c Fret saw

   d Chop saw

8 What does RCD stand for?

   a Restricted current device

   b Remote current device

   c Residual current device

   d Relenting current device

9 What does SDS stand for?

   a Slotted drill system

   b Special drill system

   c Slotted drive system

   d Standard drill system

10 What information is given on a PAT label?

   a Voltage of tool

   b Date of manufacture

   c Date of testing

   d Length of power lead

11 Recessing the head of a screw is carried out using which type of bit?

a Countersink

b Auger

c Twist

d Forstener

12 LEV refers to

a Extraction

b First aid

c Sharpening

d Metal type

13 Which legislation specifically covers hazardous substances?

a PUWER

b HASAWA

c COSHH

d RIDDOR

14 Which type of screw head is shown?

a Philips

b Torx

c Pozidriv

d Star

15 You are asked to use a PZ3 bit for a screw. What does the 3 refer to?

a The type of bit

b The size of bit

c The composition of the bit

d The length of the bit

16 Which tool would be best for cutting curves?

a Wide band saw

b Circular saw

c Chop saw

d Jigsaw

17 What would a belt sander **usually** be used for?

a Fine finishing work

b Heavy finishing work

c Forming rebates

d Forming moulds

18 Which of the following profiles would **not** incorporate a curve?

a Chamfer

b Ovolo

c Ogee

d Torus

19 Which type of bit is this?

a Auger

b Forstener

c Flat

d Countersink

20 What colour code is used for 240V?

a Pink

b Blue

c Yellow

d Red

Chapter 5
Unit 115: Produce woodworking joints

When discussing woodwork joints, it is important to consider the properties of timber and timber products and how these affect workability (how easy a particular timber is to cut and shape using tools). An understanding of how the seasoning (drying) and conversion (cutting up of a log) process, as well as an awareness of possible defects that can be found will help greatly when selecting timber and forming joints.

By reading this chapter you will know how to:

1 Select and store materials used to produce woodworking joints.

2 Select and use hand tools to produce woodworking joints.

3 Identify resources required to mark out woodworking joints.

4 Mark out woodworking joints.

5 Select and use hand tools and materials to produce basic woodworking joints.

6 Form a frame using woodworking joints.

SELECT AND STORE MATERIALS USED TO PRODUCE WOODWORKING JOINTS

TIMBER

All timber falls within two main categories, hardwood and softwood. The names imply that these materials are hard or soft, but this is not necessarily the case. The terms refer to the type of tree that the timber has been cut from. Softwood comes from coniferous trees, which are generally evergreen, have needle-like leaves and, in the case of pine trees, produce pinecones. Hardwood comes from deciduous trees. These are generally broad-leaved trees which tend to lose their leaves in the autumn, like oak (but this is not always the case). These two types of tree are different and the timber contained within has a different structure.

The following table shows different types of wood and a description of their **workability** and **durability**.

Workability

How easy types of timber are to work

Durability

How resistant types of timber are to fungal and insect attack

Wood type	Description
White wood	A type of softwood most commonly used for carcassing (studwork, joists and rafters). It is quite soft, but the knots are very hard. It is difficult to neatly work it with a chisel. This wood is non-durable and requires treatment if used in damp conditions.
Redwood	A type of softwood commonly used for joinery. It is easier to work with than white wood and is somewhat more durable, but still requires treatment if used in damp conditions.
Douglas fir	A high quality expensive softwood used for joinery and can be used 'green' (un-seasoned) for heavy construction timbers. It will require treatment for use outdoors.

Wood type	Description
Cedar 	A softwood for exterior use. It is quite soft and not particularly easy to work with. However, it is very durable and will last many years outside without treatment.
Yellow pine 	An expensive softwood highly prized for its decorative straight grain with very few knots.
Beech 	A hardwood used for good quality interior joinery and furniture. Mallets are commonly made from beech. Oak timber is nondurable, so it will go black and rot quite quickly if used outside. It works well and gives a good finish.
Oak 	A very strong hardwood, often referred to as 'the king of hardwoods'. It is very durable and is used for heavy duty timber framing, joinery and furniture. It is difficult to work with due to its hardness, although it is easier to work when green.
Ash 	A good quality hardwood, which is not very durable so is only used in dry areas. However, it will give a high quality finish, and is easy to work with.

ACTIVITY

Follow this link and plan out a wood of your choice, identifying the types of trees you would plant, how many, and why those choices. You have an area of three acres.

www.woodlandtrust.org.uk/en/planting-woodland/planting%20and-management-advice/choosing-trees/Pages/default.aspx

Wood type	Description
Sapele 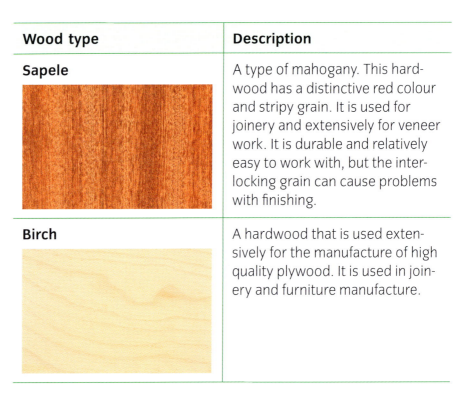	A type of mahogany. This hardwood has a distinctive red colour and stripy grain. It is used for joinery and extensively for veneer work. It is durable and relatively easy to work with, but the interlocking grain can cause problems with finishing.
Birch	A hardwood that is used extensively for the manufacture of high quality plywood. It is used in joinery and furniture manufacture.

PROCESSING OF TIMBER

Trees are not really useable until the timber inside is processed. This involves cutting up the log into useable sizes (conversion), and drying it (seasoning). Structural timber (timbers used to carry a load such as a roof or floor) are then tested for strength (stress graded) and sometimes pressure treated with preservative to increase durability.

- *Durability*. Different timber species have differing resistance to fungal and insect attack. Timbers which are resistant are known as durable while timbers that rot easily are known as non-durable. Preservatives can be applied to such timbers to increase their durability.

- *Workability*. Some timbers are easier to work with than others. Factors that affect workability can be sloping or interlocked grain, resin content, amount of knots and other defects.

TIMBER CONVERSION

A tree log can be cut up in the following ways:

INDUSTRY TIP

Pressure treated timber is where the timber has had a preservative forced into the cells of the timber in a vacuum tank.

ACTIVITY

Is there any timber in the room where you are sitting? What type is it? Why do you think that this timber been chosen?

ACTIVITY

Using www.forestry.gov.uk see if you can correctly identify trees in your local area.

Through-and-through conversion (tangential and some radial boards)

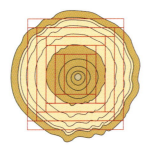

Tangential cuts (heart is boxed)

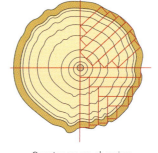

Quarter sawn, showing two different cuts (radial boards)

Boxed heart

- *Through-and-through* is the most common form of conversion. It gives an efficient use of the log – very little is wasted. The boards towards the edge of the tree are liable to become cupped in shape, this is due to the grain shape within the board. Some logs that have been cut up this way become **boules**.

- *Tangential* conversion is used for structural timbers (like floor joists) as the grain configuration makes the most of the tree ring's strength.

- *Quarter sawn* is rather wasteful, with a lot of the timber being thrown away. However, it is used for certain hardwoods such as oak because this shows up the 'figure' (medullary rays) of the timber, an attractive pattern of rays that runs radial from the centre of the tree.

- *Boxed heart* is a combination of other methods but avoiding the heart of the tree. This method is used if the heart of the tree is defective (ie rotted).

The tables below show the standard width and thickness for sawn softwood boards, and the standard lengths.

Width	Thickness	Length
75mm	16mm	1800mm
100mm	19mm	2100mm
115mm	22mm	2400mm
128mm	25mm	2700mm
138mm	32mm	3000mm
150mm	38mm	3300mm
175mm	47mm	3900mm
200mm	50mm	4200mm
225mm	65mm	4500mm
	75mm	4800mm
	100mm	5100mm
		5400mm

Boule

A log that has been sawn through-and-through and restacked into the original shape of the log

ACTIVITY

Follow this link and watch an oak log being processed:
www.youtube.com/watch?v=yigFtAuUPDE&feature=player_detailpage

INDUSTRY TIP

'Wrot' timber is timber that has been planed, and 'unwrot' timber is timber that has been planed.

INDUSTRY TIP

Planed all round (PAR) is timber which has been planed on all four sides. Planed square edge (PSE) is timber which has been planed with a square face and edge.

INDUSTRY TIP

The most common form of green timber is green oak, which is used in timber frame construction.

DRYING

When a tree is cut down, it is full of water and sap (this can be up to 200% of the dry weight). This needs to be removed before the timber is useable through a process known as drying or seasoning. Green timber (un-seasoned) is heavy, usually difficult to work and is prone to rot or insect attack. Green timber will also move after it has been used resulting in warping and splitting. However, some timbers, such as oak, can be used 'green', and the splitting and warping process is deemed to be attractive.

Drying or seasoning can be done naturally or artificially. The aim of drying timber is to ensure the finished product is suitable for its intended task. The main reasons being:

- To reduce the moisture content to below 20%, this is the threshold of dry rot.
- To reduce the likelihood of shrinkage.
- To enable good surface finish to be achieved both by machine tools and hand.
- To help with decorative finish that will be applied to the finished product.

The natural process is known as air seasoning (drying) which is where the freshly cut timbers are placed under a cover but air is allowed to pass through the stack. Slowly the timber dries out.

Moisture meters are used to measure moisture in timber

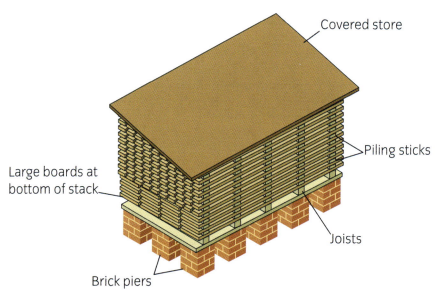

Air is able to pass through the boards when timber is correctly stacked with piling sticks

Case hardening

A defect caused by timber being kiln dried too rapidly, leaving the outside dry but the centre still wet

This results in stable material, but it can take a long time. Modern kiln drying involves the use of ovens and is much quicker but the timber can sometimes have some serious seasoning defects, such as **case hardening**, if care isn't taken during the drying process.

A kiln oven used for drying

Shrinkage

Shrinkage of timber is twice as much along line A as it is along line B, in the image below on the left. This causes timber to change shape in the following ways in the image on the right.

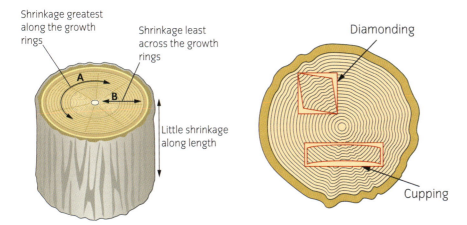

Growth rings

Every year a tree grows it adds a ring and increases in diameter. An annual ring consists of winter growth and summer growth. In softwoods, the winter growth is the darker part and is stronger than the summer growth, therefore slower grown softwoods are stronger because there are more rings in a section. The bark of the tree has a number of layers. Each layer is made up of cells which help the tree protect itself and supply it with nutrients. The images on the next page show you the different structure of tree bark.

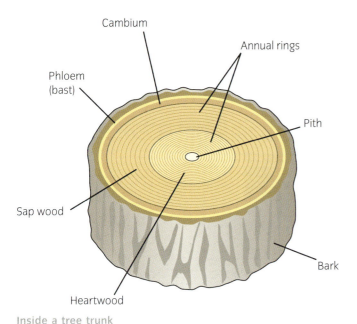

Inside a tree trunk

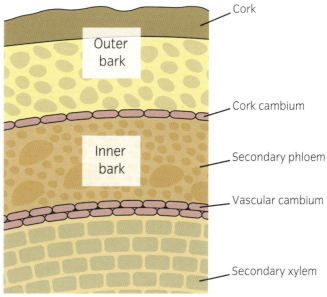

Structures within the layers of bark

STRESS GRADING

Stress grading is usually done by passing the timber through rollers under pressure. The deflection (bend) the timber shows is measured. The greater the deflection, the weaker the timber is, and the lower the grade of the timber. The grade is stamped on the timber which is either C16 (lower grade structural timber) or C24 (higher grade structural timber).

C16 and C24 stamps on timber

OUR HOUSE

Timber floorboards have shrunk after being laid in a house. You have been asked to list possible reasons for this and also to find a solution so it does not happen again. For help with this, take a look at the boards in 'Our House'.

BOARD MATERIALS

Timber can be further processed into board materials. Common types of manufactured board are:

Board type	Description
MDF 	MDF stands for medium density fibreboard. It is a very common material made by breaking the timber down into fibres using water. These fibres are then dried and pressed into boards using adhesive. MDF produces a lot of dust when cut.
Chipboard 	Chipboard is made from small chips of wood, bonded (stuck) together with adhesive. It is cheap, but quite brittle. Chipboard can be coloured green to show that it has been treated to be moisture resistant, or coloured red to show that it is fire resistant.
OSB 	Orientated strand board is similar to chipboard but has much larger chips which makes the resulting board much stronger. OSB is often used for structural purposes.
Plywood 	Plywood is made from many thin layers (or laminates) stuck together at right angles to each other, making it strong. Plywood usually has an odd number of plies or layers which makes the board stable.  Layers within plywood Plywood is stronger than chipboard. The quality of the resulting product depends on the quality of the timber and adhesives used. WBP (water/weather boil proof) uses an external grade adhesive that enables it to be used inside or outside. Moisture resistant (MR) plywood has reduced resistance to water. INT (interior grade) is only suitable for internal use, it will not stand up to damp conditions.

The following are other types of plywood:

Plywood type	Description
Marine ply	A very high grade plywood with an excellent resistance to delaminating (coming apart) in extreme conditions.
Birch ply	A high quality plywood used for furniture.
Flexi ply	A soft, flexible plywood used for forming shapes.

The following are other types of laminated board:

Laminated board type	Size and description
Blockboard	Strips of board are up to 25mm wide. Good quality hardwood veneers are sometimes used for this board.
Laminboard	Strips of board are 7–8mm wide. This size produces a better quality board.
Battenboard	Strips of board can be up to 75mm wide. This size produces a poorer quality board.

TIMBER DEFECTS

The following are common natural defects that occur in timber:

Defect	Description
Heart shake 	A defect that occurs from the centre of the tree (in the heartwood), sometimes associated with decay.
Star shake 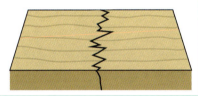	Shakes in the timber radiating from the centre. A star shake is a collection of two or more heart shakes.
Thunder shake 	A crack across the grain. Also known as an upset.
Cup shake	When the growth rings start to come apart, causing a shake that runs the circumference of the growth rings. Also known as ring shake.

Defect	Description
Knots 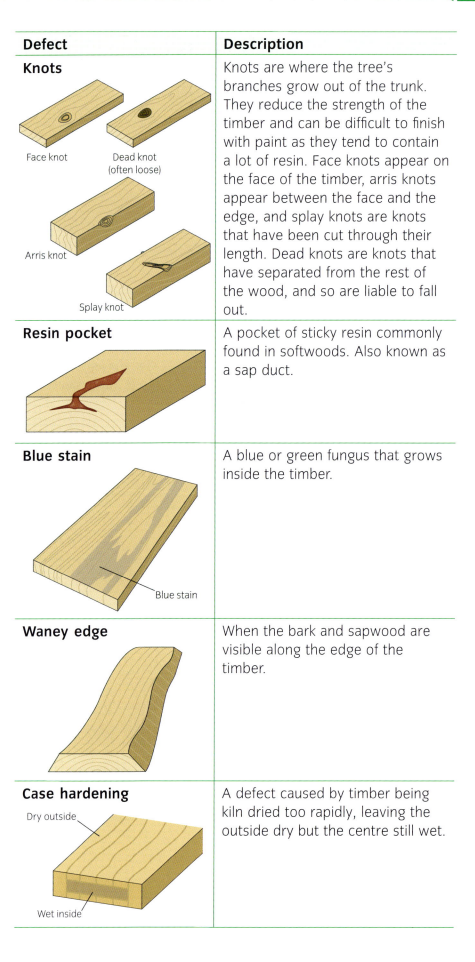 Face knot Dead knot (often loose) Arris knot Splay knot	Knots are where the tree's branches grow out of the trunk. They reduce the strength of the timber and can be difficult to finish with paint as they tend to contain a lot of resin. Face knots appear on the face of the timber, arris knots appear between the face and the edge, and splay knots are knots that have been cut through their length. Dead knots are knots that have separated from the rest of the wood, and so are liable to fall out.
Resin pocket	A pocket of sticky resin commonly found in softwoods. Also known as a sap duct.
Blue stain Blue stain	A blue or green fungus that grows inside the timber.
Waney edge	When the bark and sapwood are visible along the edge of the timber.
Case hardening Dry outside Wet inside	A defect caused by timber being kiln dried too rapidly, leaving the outside dry but the centre still wet.

Defect	Description
End splitting	Shakes due to loss of moisture at the end of a plank.
Collapse	Where the cells in the timber are dried out too quickly resulting in the destruction of the timber's structure.
Twisting	A defect where the timber distorts in such a way that the timber becomes shaped like a propeller.
Bowing	A curvature (where the wood curves or bends) along the board's face (the flat of the timber) from one end to the other.
Springing	A curvature along the board's edge, from one end to the other, but the face is still flat.
Cupping	Where timber 'cups' over the face of the board. Wide boards that are cut tangentially will usually cup over time.

Enemies of timber

Timber can be rendered unusable by rot and by insect attack. There are two main types of rot caused by fungus: dry rot and wet rot. Dry rot will stop as soon as the source of dampness is stopped, and only badly damaged parts require replacement. Dry rot is much more serious than wet rot and it requires specialist treatment which involves the removal of unaffected timber near to the attack, which is either disposed of in a controlled manner to prevent spread of spores or it is burned. There are a few insects that attack timber: the most common is the furniture beetle, which attacks both softwood and hardwood. The image below shows a wood boring insect's life cycle.

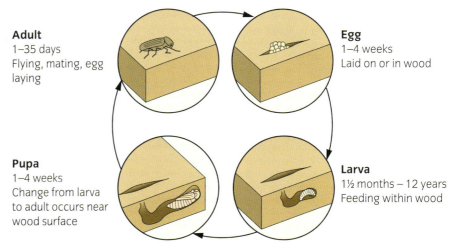

Adult
1–35 days
Flying, mating, egg laying

Egg
1–4 weeks
Laid on or in wood

Pupa
1–4 weeks
Change from larva to adult occurs near wood surface

Larva
1½ months – 12 years
Feeding within wood

Life cycle of wood-boring insect

STORAGE OF TIMBER

Timber and board materials should be stored in the dry, flat and out of **twist**; sheet materials are best stored flat on bearers but can be stored short-term on upright racks. Incorrect storage will result in the material being damaged and could render it unusable. It can be damaged by water, ultra violet light (UV) or be damaged by work operations (spillages, collisions or other accidents). Materials could also be liable to theft.

Timber stored upright

Twist

A seasoning defect where there is a spiral or corkscrew distortion in a longitudal direction of the board

Incorrect storage – timber should be stored in the dry and away from UV

Very dry timber used outside is likely to swell, and wet timber used inside will shrink causing cracks. If possible, it is good to get timber acclimatised to where it will be used, for instance storing floorboards in the room where they will be laid; this is commonly termed 'second seasoning'.

SUSTAINABILITY

The planet we live on is a fixed size so resources are limited. Materials such as timber are sustainable and renewable as long as it is grown in a managed way. This means that new trees are planted as other trees are cut down. Trees absorb carbon dioxide as they grow through **photosynthesis**, are nice to look at, and are cheaper to build with than many other materials. For more information on sustainability refer to Chapter 2, pages 92–97.

TYPES OF WOODWORKING JOINT

A wide variety of methods can be used to join timber, depending on the end use of the job, cost and desired quality of finish.

OUR HOUSE

'Our House' contains a number of woodwork joints. When reading through this chapter, see if you can find each joint in the house as well.

COMMON JOINTS AND THEIR USES

Butt joint

Butt joints are the simplest joint, where one piece butts up against another and is fixed with nails or screws. These joints are usually used for constructing studwork. The nails or screws used to fix the joints can be 'dovetailed' (skewed) in order to increase the strength of the joint.

Photosynthesis

The process by which plants convert sunlight into food. Oxygen is produced as a by-product

Roof trusses are made with butt joints fixed with nail plates

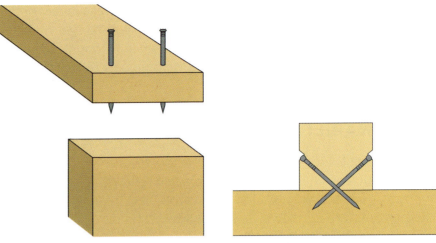

Butt joints in stud with nails

Dovetailed butt joint

Roofing joints

Roof rafters have these joints. Vertical cuts are called 'plumb cuts' and horizontal cuts are called 'seat cuts', and a plumb and seat cut together at the bottom of the rafter is known as a 'birdsmouth cut'.

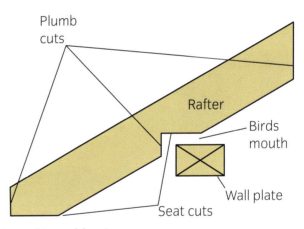

Cuts and joints used in roof framing

Halving joint

A halving joint is where half of each of the two pieces of timber being joined is removed, so that the two boards join together flush with one another. They are commonly used in jointing timber lengthways or as simple frame (corner) joints. They are not strong joints and have to be fixed in place using nails, screws or adhesive.

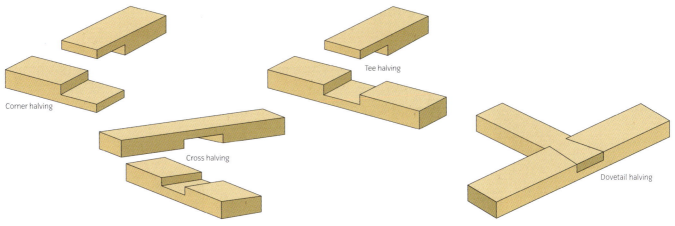

Four types of halving joints

Bridle joint

The bridle joint is a simple framing joint. The timber is divided into three equal parts in its thickness, with the centre piece removed on one part and the two outsides removed on the other. This joint is stronger than a halving joint as the glue area (contact face) is increased and, compared with the halving joint, can only be taken apart in one direction. These can be secured using nails, screws or adhesive.

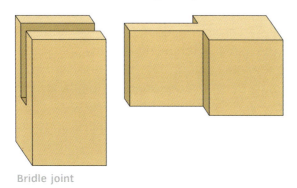

Bridle joint

Housing joint

The housing joint is a commonly used joint for wide material, eg making door linings and shelving units. Variations of the housing joint can be used to suit varying circumstance, ie through, stopped and tongued (also known as barefaced) housing joints are used for door lining, shelving/cupboard and stair construction. Again this joint can be secured with nails, screws or adhesive.

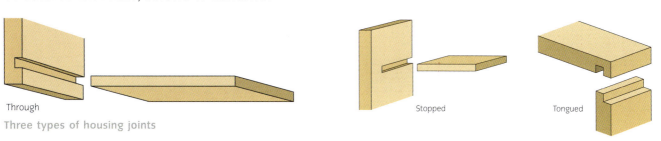

Three types of housing joints

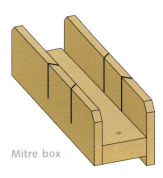

Mitre box

Mitre joint

The mitre joint is commonly used to join moulded finishes at internal and external corners such as skirting or architraves. It is fixed in place with nails or pins and adhesive is required. They can be cut by machine using a chop saw or by hand using a mitre box and shooting board (used to easily trim the mitre with a plane).

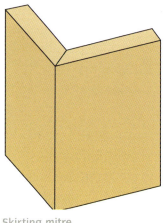

Skirting mitre

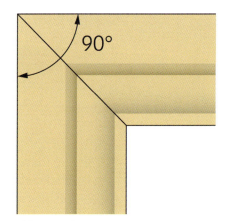

Architrave mitre

Mortice and tenon joint

Mortice and tenon joints are the most common joint used in joinery construction. It is a very strong joint, available in many variations depending on the joinery item being constructed. The mortice is a slot, known as the female part, and the tenon fits into the slot, known as the male part. It is the principle joint used in window and door frame, door and sash construction. These can be secured using wedges, adhesive and sometimes screws nails or dowels if used in the construction of frames.

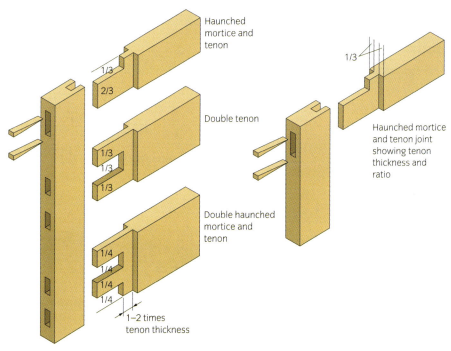

Mortice and tenon joints

Dovetail joint

A strong, high quality joint, the dovetail joint is traditionally used in the construction of drawers. The pitch (angle) of dovetails resist the joint being pulled apart under load. These joints are held together using adhesive alone, no additional fixings should be required. The pitch or slope of the dovetail should be 1:6 for softwood and 1:8 for hardwoods. Some people use the average of these two pitches (1:7) for all dovetails, whether in hardwood or softwood.

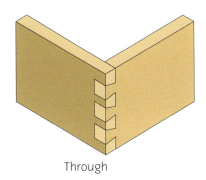

Through   Lapped

Two types of dovetail joints

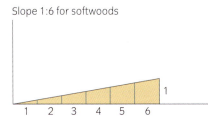

Slope 1:6 for softwoods

Slope 1:8 for hardwoods

Pitches for hard and softwood

Lengthening joint

Occasionally timber will require jointing in its length. Lengthening joints can be categorised as either structural or non-structural. The most common non-structural lengthening joint is called a heading joint. An example of this would be where a floor or skirting board is jointed in its length. This can be simply square or it can be improved by being 'splayed'. A structural lengthening joint would be required where load bearing components require jointing. An example of this would be jointing a ridge board or purlin used in roof construction. A scarf joint is a type of non-structural joint used for timber and veneers and commonly found on a ridge board.

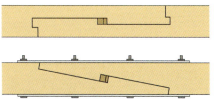

Structural scarf joints

Scarf joint in ridge board at the top of a hand cut roof

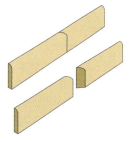

Splayed heading joint in skirting board

Square heading joints in floor boards supported on joists

Edge joint

Edge joints are used to make timber boards wider. They can be 'rubbed' (where glue is applied and the two pieces rubbed together to ensure good coverage of the joint), tongued, biscuited or doweled to increase strength. The tongue, biscuit and dowel all increase the interconnection of the joint.

The images below show the following joints:

- *Biscuit joint.* This joint is made using a biscuit jointer power tool.
- *Loose tongue.* A groove is placed in each piece to be joined and a plywood tongue joins the two together.
- *Dowels.* Used to join boards on the edge or (as shown in the kitchen unit image) in the corners. Biscuits can also be used in this way.

Always ensure that when gluing multiple boards together you place the growth rings (see page 197) alternatively up and down to reduce distortion.

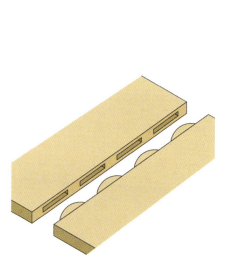

Edge joint with biscuits

Edge joint with a loose tongue

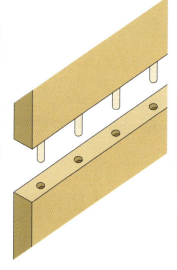

Edge joint with dowels

A kitchen unit jointed with dowels

DOOR AND FRAME COMPONENTS

It is important to know the names of the components that make up a frame before you start assembling one yourself. A door frame is shown below. The parts of a sash (the opening part of a window) also have the same names, ie stiles and rails.

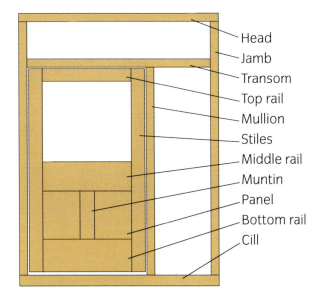

- Head
- Jamb
- Transom
- Top rail
- Mullion
- Stiles
- Middle rail
- Muntin
- Panel
- Bottom rail
- Cill

MARK OUT WOODWORKING JOINTS

Now that you know the materials, the tools (from Chapter 3) and some joints, you are ready to start making a frame. Joinery items are often constructed in the form of frames and to assist in this process it is good practice to draw a rod. This is a full-size drawing of the item to be made. On this drawing, details such as the joint proportions and the distances between shoulders can be determined.

DRAWING A ROD

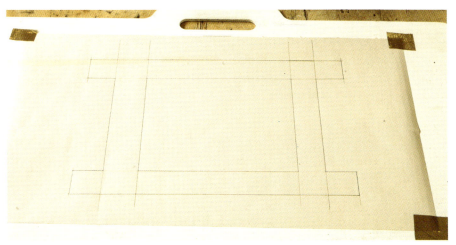

A rod for a frame

Using a working drawing as a guide, overall measurements are transferred to the rod. This will include the length and width of the frame, along with sizes of the materials to be used. (Rebates, mouldings and other details can be added at this stage allowing for clearance gaps around sashes or doors if so required.)

It is good practice to **hatch** parts of the rod that indicate a section. Remember to add any other details such as job name, or date. Once this has been completed, this rod will be used to mark out the timber and fill out a cutting list.

Hatch

Marking waste timber with diagonal lines. This prevents mistakes, as only the hatched area is removed

CUTTING LISTS

These are used to record the components required for a given joinery task. Included on the list are details such as name and number of the components, sizes (width, thickness and length) and comments. Additionally there may be extra columns giving the 'ex' or 'nominal' size – this refers to **sawn size**. This is often 5 or 6mm larger than the finished size.

Sawn size

The timber size before planing

Cutting list								
Component	Species	Number	Length	Nominal width	Nominal thickness	Finished width	Finished thickness	Comments
Stiles	R/wood	2	2050	100	50	95	45	
Top rail	R/wood	1	800	200	50	195	45	
Mid and bottom rail	R/wood	2	800	200	50	195	45	

Produce a cutting list for this item of joinery:

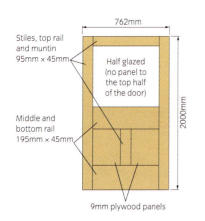

Stiles, top rail and muntin 95mm × 45mm

Half glazed (no panel to the top half of the door)

Middle and bottom rail 195mm × 45mm

762mm

2000mm

9mm plywood panels

'Ex' is an abbreviation of 'extracted from' and 'nominal' means before anything is taken away. Both these terms are commonly used to describe the sawn sizes prior to planing.

MARKING OUT A SIMPLE FRAME

The following step by steps show how to mark out a frame using a rod. The same process is used for marking out a door; use the stiles where jambs are mentioned.

STEP 1 First select the face side and edge of all the pieces. The face side and edge are the best sides, any defects will be kept to parts of the frame least seen, eg the back of a frame or outside face, or where a moulding or rebate would remove the defect.

STEP 2 Make sure you know which piece of wood goes where. Lay them out around the rod if required, and write on the back if you wish to do so.

STEP 3 Lay one of the **jambs** onto the rod. Mark out where the shoulders should be cut, and where mortices will be created. This first piece is known as a pattern.

STEP 4 Put the two jambs together as a pair (a left hand and right hand jamb) face to face and square lines across from one to the other. This ensures the two pieces are exactly the same length – the frame will be an even size.

STEP 5 Repeat the process for all the other pieces, mark out a pattern first from the rod, then use this pattern to mark out any other pieces that have similar shoulder dimensions. Pieces with mortices can be marked in a similar way.

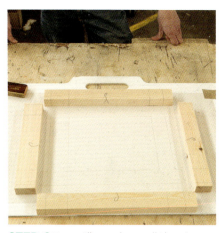

STEP 6 You will now have all the pieces marked out from the rod with positions of the joints.

When marking out, traditionally the components with the mortices are marked out first. These can then be given to the machinist who will mortice these while the jambs are being marked out. This speeds up production and makes the most efficient use of time.

Jambs

The vertical (upright) parts of a frame

INDUSTRY TIP

It is best to use a 2H pencil for marking out as it will leave a nice sharp line. HB is too soft and will leave a wide line leading to inaccuracy. It is also likely to smudge, leaving marks all over your wood!

Marking out a mortice

Determine the position of the mortice. It is usually near the centre of the timber section, but can be moved if required. The closer to the centre the mortice is, the stronger the joint. On smaller stock the mortice is usually 1/3 the thickness of the timber but will be dictated by the size of mortice chisel available, and the mortice gauge will be set to the chisel size – '12mm' chisels can vary in actual size and be from 11mm up to 13mm.

Marking out a mortice

The following step by step guide shows you how to centralise the mortice on the timber section.

STEP 1 Set the mortice gauge to the chisel size.

STEP 2 Place on the timber and make two holes.

STEP 3 Turn the timber over and repeat. If the holes fall in the same place then the mortice will be in the centre.

STEP 4 If the holes do not line up, adjust the gauge to fall between the holes you have made and the gauge will be in the centre.

When gauging, roll the pins into the timber at one end of the mortice/tenon, push or pull the gauge from the other end until the pins fall into the two holes. This automatically stops the gauge lines from 'over running' the joint, damaging the inside face and showing after it is assembled.

STEP 1 Mark the mortice. Always mark out from the face.

STEP 2 Make a hole at the end of the line you are going to mark, then gauge into the hole.

Hatch the mortice

Use a pencil to hatch the mortice, ie marking it with diagonal lines. This will ensure that you only cut out that mortice area and so avoid mistakes. The hatched area is the part that you will remove. Marking out a bridle or halving joint follows a similar process.

INDUSTRY TIP

Haunches are used at the ends of frames to enable the joint to be wedged and prevent twisting in the rail.

Hatching the mortice

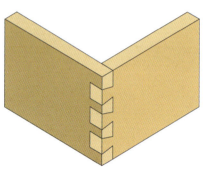

Interlocking pattern in a dovetail

Marking out a dovetail

Dovetails are made with an interlocking pattern. The pitch of 1:6 is used for softwood and 1:8 for hardwood. Some people use 1:7 for everything. For more information on dovetail joints, turn to page 209.

The following step by steps show you how to mark out a dovetail joint.

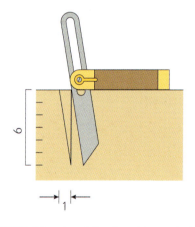

STEP 1 Set the sliding bevel to mark the pitch.

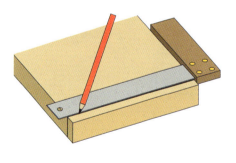

STEP 2 Mark the shoulder lines.

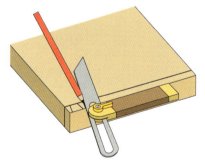

STEP 3 Mark the dovetails on the timber.

INDUSTRY TIP

Measure twice, cut once! Take some time to check your measurements and marking out, once it's cut you can't glue it back on.

When producing joinery work it is good practice to mark out all pieces at once, then cut the joints. However, with dovetails the dovetail is cut, and then the pins are marked out from the dovetails.

The following step by steps show you how to cut out and assemble a dovetail joint, after you have marked out the dovetails.

STEP 1 Place the timber in a vice and cut out the dovetails that you have marked on the tail board.

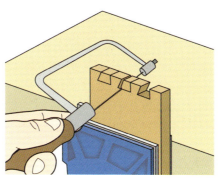

STEP 2 Use a coping saw to remove the waste between the dovetails.

STEP 3 Using a bevel edged chisel, pare to the shoulder lines on the tail board.

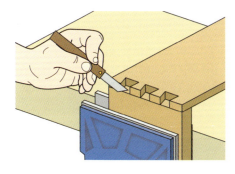

STEP 4 Use the dovetails to mark out on the mating piece (the pin board). Use a pencil for softwood and a marking knife for hardwood.

STEP 5 Cut out the pins on the pin board.

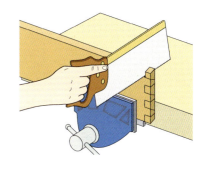

STEP 6 Saw off the corner waste at the ends of the boards.

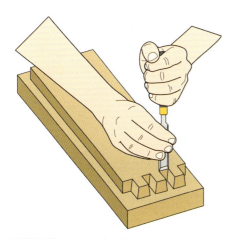

STEP 7 Using a chisel, pare to the shoulder lines on the pin board.

STEP 8 Assemble the dovetail joint using a mallet.

PPE safety boots

ACTIVITY

Follow the process for cutting a tenon with your own piece of wood.

FORM A FRAME USING WOODWORKING JOINTS

Now your frame is marked out, it's time to cut the joints.

Safety first. The workshop has many potential hazards, from items falling on feet to cuts from tools to hazardous dust. Appropriate PPE should be worn, as stated in the risk assessment that will have been put into place. A mask will prevent dust inhalation, safety boots will protect your feet from falling objects and ear defenders will protect your ears from excessive noise. You can find more information on health and safety legislation and PPE in Chapter 1.

RIPPING

The term ripping refers to cutting down the grain. When cutting a halving joint, bridle or tenon the first job is to rip cut the cheeks. The cheeks are the sides of the tenon, bridle or halving.

CUTTING A TENON

The following process explains how to rip cut tenon cheeks. This process is the same for a bridle or halving joint.

1 Start sawing. Use your thumb as a guide. As soon as the saw starts, take your thumb away, and be very careful to keep it away from the teeth. Try to use the weight of the saw rather than force it to cut. Notice the cut is on the waste side, but following the line.

Using your thumb as a guide when sawing

2 Cut a depth of about 5mm.
3 Lean the timber over to about 45°, cut a diagonal from corner to corner. Notice that you will be able to see both ends of the cut and, with some practice, be able to cut straight.
4 Turn the piece over, repeat the process.
5 Now finish the cut. The tenon is now ready for the shoulders to be cut.

INDUSTRY TIP

Tenon saws are used to cut shoulders.

Finishing the cut

CROSSCUTTING

This is cutting across the grain, and will form the shoulders of the joint. The images below show the shoulders of different types of joint.

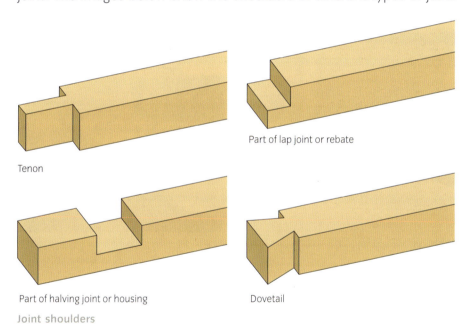

Tenon

Part of lap joint or rebate

Part of halving joint or housing

Dovetail

Joint shoulders

The following step by step guide shows you how to crosscut timber. To begin place the timber in the bench vice, taking care not to have the cut too close to the metal as this could damage the tenon saw. This will enable you to hold the timber properly and safely. As with ripping, start the saw using your thumb as a guide.

STEP 1 Take great care and take your thumb away as soon as the cut is started.

STEP 2 Finish the cut. Be careful not to overcut past the gauge line because this will weaken your tenon.

HOUSING

This cut is the same for a housing or bridle joint.

STEP 1 Crosscut down to the gauge lines using a tenon saw. If there is a large amount to remove, or there is a big knot in the middle, several other cuts can be made.

STEP 2 Using a bevel edged chisel, cut upwards forming a slope. Don't go right through because the timber is liable to split. Turn the timber around, and repeat the process. You will end up with a little 'roof' shape in the middle of the housing.

STEP 3 Clean out the middle, again taking care not to go right through the joint with your chisel.

STEP 4 The finished housing joint.

MORTICING

Morticing should be done using a mortice chisel. A normal bevel edged chisel is thin, using the wrong chisel reduces the accuracy of the joint and also means the chisel might bend or break. Your workpiece will require good cramping. It can be cramped to the bench, above a leg if possible. The following step by steps show how to perform morticing.

A set of mortice chisels

STEP 1 Position your chisel in the centre so that you are ready to start morticing.

STEP 2 Work your way out, ensuring the waste is ejected each time. Keep away from the end of the joint – you will clean these out later.

STEP 3 Repeat this process until you are about halfway through the timber, then turn it over. Repeat process. Do not go right through the timber with your chisel, it will split the wood. Clean out the ends of the joint. Clear out the mortice with a blunt instrument such as a rule or the end of a combination square.

An allowance can be made for wedges if required.

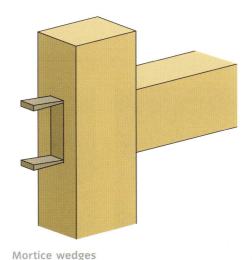

Mortice wedges

ASSEMBLING THE FRAME

Once the joints have been cut it's time to dry fit the frame. Any adjustments will be made at this stage. A set square can be placed on the face of the joints to check they are flat and not twisted. When all the joints are correctly fitting together, you are ready to assemble your frame.

You should assemble the frame dry to check everything is fine (the joints should fit well and the frame is not twisted or out of square), and check you have everything you need before putting any adhesive on the joints. You will need **sash cramps**, some small blocks of wood, wedges, screws or nails as required, and adhesive. There are many adhesives available – this is a list of those commonly used.

- *PVA*. Polyvinyl Acetate. Commonly referred to as 'white glue'. A resin dissolved in water. As the water evaporates the glue dries (goes off). Available in interior and exterior grades. It has good gap-filling properties and provides a strong permanent bond. It dries clear, but if not cleaned up properly can leave marks that become visible when the product is varnished or stained.

- *PU*. Polyurethane. Commonly referred to as 'foam glue'. A yellow-brown resin that foams upon contact with air. It forms a strong water resistant bond and has excellent gap filling properties. This adhesive can be used to join damp timbers.

- *UF*. Urea-formaldehyde. Commonly referred to as powdered resin glue. This adhesive is mixed with water before use. A chemical reaction causes it to set after a few hours depending on the surrounding temperature. It forms a strong water resistant bond. It is used for high quality work and does not stain the timber being joined.

The following step by step guide shows you how to assemble a frame.

Sash cramp

A cramp that consists of a steel bar with an adjustable 'shoe' at one end and a fixed saw at the other. It is used when assembling work to pull up joints tight

INDUSTRY TIP

Joints should fit together with a 'snug' fit. This means the joints are not tight, if you are having to bash away with a mallet to get the joint together then it's probably too tight! Tight joints are difficult to assemble and usually result in a frame that is out of square or twisted. Loose joints are weak and unsightly.

ACTIVITY

Using the glue you have to hand, form a cross halving joint, glue it together and then when fully set see how easy it is to break the joint. A good glue joint will hold together and the timber around the joint will break apart.

INDUSTRY TIP

When checking for square, if both diagonals are the same the frame is square. If they differ, one cramp can be positioned towards the long diagonal so as to apply a little pressure to this and shorten it. Keep making adjustments until the diagonal lengths are equal.

STEP 1 When the frame is ready to assemble, place the parts in the correct position. Make sure that you have the bearers and cramps on hand and that they are placed level and out of twist.

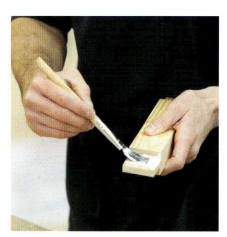

STEP 2 Apply adhesive to the joints, and quickly assemble. Apply just the right amount of adhesive, enough to bond the joints together but not so much that it squirts out everywhere.

STEP 3 Cramp the frame together. Check for square using a squaring rod to measure both diagonals, and check for twist by checking if the frame sits flat on the levelled bearers. If it does, then it is not twisted.

STEP 4 The wedges should not be driven until it is square and flat, as once they are driven they can't be altered. When correct drive the fixings using a hammer (if using wedges the outside ones should be driven first followed by the inside). Notice the block of wood used with the cramps to prevent damage.

STEP 5 The cramps can be removed when the joints are secured. If wedged or screwed, the job can be released from the cramps straight away. If the joints are held by adhesive alone it may need some time to set ('go off'). After the frame has been wedged, excess glue can be removed, with a damp rag and a chisel.

STEP 6 Trim the joints. Cut off or trim the projecting parts of the frame (horns) as required.

INDUSTRY TIP

Checking for square and twist is a very important part of this assembly procedure.

INDUSTRY TIP

The amount of time the adhesive will need to set will depend on the adhesive used, the glue joint thickness and the temperature and humidity of the environment.

STEP 7 Plane the faces off using a smoothing plane. Plane the back of the frame first, make sure the joints are flat (flush) don't take off too much as the frame will become too thin. Plane the front then the edges.

INDUSTRY TIP

Beware of damaging the frame with the front (toe) of the plane when planing the faces.

INDUSTRY TIP

Check the plane for sharpness and set then try the plane out on a scrap piece of timber before planing the faces of the frame.

MAINTAINING A TIDY WORKSHOP

Don't forget to clean up after yourself upon completion. Sawdust, shavings and offcuts must be cleared up and disposed of safely. Unused timber can be returned to the store for future use. A clean work area is a safe work area! After using your tools always check them for damage, clean and sharpen them as necessary and store them in the correct place. Well maintained tools will always work better and produce better results than poorly maintained tools.

Case Study: Akeem

Akeem has been asked by his supervisor to make some doors and windows for a local building project, a barn conversion. There are internal and external doors, the external doors being matchboarded with tongue and groove. The building is in an exposed position with the sea nearby. The clients are watching the budget very carefully, so even though they wish for a good durable finish to the exterior of the building, they are willing to reduce costs with the interior.

- What species of timber should Akeem select for the internal doors and why?

- Which species of timber should he select for the external doors and why?

- When selecting the tongue and groove for the external doors, which grain pattern should Akeem look for to ensure greatest stability?

- Make a sketch of the grain pattern he should use, and where this type of timber would appear on a log converted by the through-and-through method.

Work through the following questions to check your learning.

1 Which **one** of the following is a growth defect in timber?

a Knot

b Twist

c Shake

d Springing

2 Which **one** of the following is a seasoning defect?

a Twist

b Resin duct

c Knot

d Upset

3 Which **one** of the following is a softwood?

a Oak

b Ash

c Sapele

d Pine

4 C16 is a timber classification. What does it refer to?

a Strength

b Length

c Width

d Weight

5 What is PVA glue commonly known as?

a Foam glue

b White glue

c Black glue

d Water glue

6 What type of saw would be used for cutting joint shoulders?

a Rip

b Panel

c Tenon

d Pad

7 Which **one** of the following pitches would be used when dovetailing in softwood?

a 1:5

b 1:6

c 1:7

d 1:8

8 In relation to plywood, what does MR stand for?

a Moisture mouldy

b More resistant

c More resilliant

d Moisture resistant

9 A scarf joint is likely to be found on a

a Floorboard

b Table

c Drawer

d Ridge board

10 Which of the following joints is commonly wedged?

a Bridle

b Halving

c Tongued housing

d Mortice and tenon

11 Cutting down along the grain is referred to as

a Ripping

b Crosscutting

c Mitring

d Morticing

12 Plywood is manufactured from

a Laminates

b Chips

c Fibres

d Dust

13 Architraves are joined using which joints?

a Butt

b Mitre

c Heading

d Bridle

14 How can a frame be best checked for square?

a Compare with another frame

b Lay on the floor

c Check the diagonals

d Use winding sticks

15 Wedges are best driven using a

a Mallet

b Hammer

c Punch

d Cramp

16 How should joints fit before assembly?

a Snug

b Hammer tight

c Loose

d Mallet tight

17 Using a levelled bench and bearers during assembly will help prevent a frame becoming

a Out of square

b Too big

c Twisted

d Too small

18 The best type of pencil to use for marking out joinery is

a 2H

b 2B

c HB

d BH

19 Jambs should be marked

a Singly

b In pairs

c Before marking the faces

d Before marking the face edges

20 Board materials are best stored in which way?

a Flat on bearers

b In the open

c At 45°

d Against a wall

TEST YOUR KNOWLEDGE ANSWERS

Chapter 1: Unit 201

1 c Risk assessment
2 d Blue circle
3 b Oxygen
4 a CO_2
5 b Control of Substances Hazardous to Health (COSHH) Regulations 2002
6 c 75°
7 c Glasses, hearing protection and dust mask
8 d Respirator
9 a 410V
10 b 80dB (a)

Chapter 2: Unit 101

1 c 15m
2 a Open to interpretation
3 b A section through a part of the structure
4 a Strip
5 c Raft
6 c Damp proof course
7 d Polystyrene
8 d Cement
9 b 10°
10 a Foundations

Chapter 3: Unit 113

1 b 2–8m
2 c 90°
3 a Sliding bevel
4 b 150mm steel rule
5 b Down the grain
6 b 8–10
7 c Saw guard

8 c Sole
9 d Smoothing
10 d Compass
11 b 6–13mm
12 d Bevel edged
13 d At the end of its shaft
14 a From both sides
15 c 30°
16 d With softened corners
17 a Slip
18 b 8mm
19 c Triangular
20 b 5mm

Chapter 4: Unit 114

1 b Carbide
2 b Measure of grit
3 d Router
4 b 110V
5 a Circular saw
6 c Positive
7 d Chop saw
8 c Residual current device
9 c Slotted drive system
10 c Date of testing
11 a Countersink
12 a Extraction
13 c COSHH
14 c Pozidriv
15 b The size of the bit
16 d Jigsaw
17 b Heavy finishing work
18 a Chamfer

19 c Flat

20 b Blue

Chapter 5: Unit 115

1 a Knot

2 a Twist

3 d Pine

4 a Strength

5 b White glue

6 c Tenon

7 b 1:6

8 d Moisture resistant

9 d Ridge board

10 d Mortice and tenon

11 a Ripping

12 a Laminates

13 b Mitre

14 c Check the diagonals

15 b Hammer

16 a Snug

17 c Twisted

18 a 2H

19 b In pairs

20 a Flat on bearers

INDEX